The Nature of Consciousness

First Edition

Karl Sipfle

Copyright © 2018 by Karl Sipfle. All rights reserved.

To Liz, and to Aristotle, Descartes, and Leibniz. Also, to Vivaldi and Goethe, who taught me that even the exquisite favors boldness over perfection.

K.S.

Contents

1. Consciousness is a Real, Natural Phenomenon 9
2. Defining terms: Levels of Consciousness 12
3. Kinds of Consciousness 17
4. Emotional Consciousness is the Foundation 25
5. Clues from Brains 30
6. Who is Doing the Feeling? 33
7. Real, Natural Phenomena Are Described by Physics and Math .. 34
8. The Consciousness Field and The Senton 38
9. Where is the Site of Action in the Brain? 80
10. From Association to Thinking 121
11. What It is Like to Be a Human 123
12. Language 132
13. Brain Structures for Higher Consciousness 133
14. Attention, Will, Willful Attention, and Attention to Will 136
15. Self and the Unity of "Me" 138
16. Destiny 145
17. Why Consciousness? 146
18. Artificial Consciousness and Trans-Human Fused Consciousness 148
19. Conclusions 153

Appendix I: Reality, Existence, Universes 156
Appendix II: Brief Summary of Our Universe 167
Appendix III: Neural Module Architecture 170
Appendix IV: Thalamocortical Attention Circuit 171
Appendix V: Concluding Speculations 177
Appendix VI: Philosophical Note 186

Preface

This book was begun in 2017 with the realization that most readers will have ready access to the Internet. It does not attempt to include tutorials on all established knowledge to which it refers. Neither does it seek to conform to the models of authors past; the problems discussed herein have been pondered and discussed for centuries by many.

Much work has been done on the biological, neural level. Work has been done to understand cognitive higher consciousness, and a bit on the importance and contribution of value, assuming the most basic mechanisms for mind are in place. In this book we look at those basic mechanisms, the missing explanatory glue from the physical to value and consciousness.

In this book we take a journey through the various levels of consciousness, on a mission to understand where it ultimately comes from. Important concepts are repeated in various sections. It is probably most effective for the reader to skim the whole book once and then re-read it more carefully.

The author is eager to improve this crude bundling of notes in a subsequent edition. Kindly send all comments to ksipfle@umich.edu. While the writing begs for improvement, comments on the ideas or means of explanation are of special interest.

# INTRODUCTION

1. Consciousness is a Real, Natural Phenomenon

Consciousness is a natural phenomenon. We know that it happens, and it happens in our natural world.

Many phenomena over time have been seen as magical and supernatural. Inevitably, orderly Nature has been discovered to be the source of them all. Much as with life itself in earlier times, this book proposes that consciousness is simply the latest "magical" mystery and is not supernatural or foreign to our ways of scientific understanding, but natural, and resulting from laws and facts of physics. As with life itself, these laws and facts are probably similar to and connected to the laws and facts that we already know. In fact, one of the most reliable rules of thumb about this universe, underlaid by its nature, is that that which we see repeatedly, we are likely to see again.

We propose and expect that we can theorize about the nature of consciousness with some specificity, and in this book, we do just that.

Much has been written on consciousness, with attempts to define many terms. In this book we take a clean-slate approach and deliberately minimize references to extant terms and theories that may suggest other than what the author intends or what is so.

Some readers may hold concerns about whether the nature of consciousness is a puzzle so different from those that have come before, as to make it resistant to analysis and discovery. In particular, consciousness may seem to be foreign to space or time- somehow more so than, say, known electrical phenomena. These readers may want to first read Appendix I.

This book makes the following major predictions

- Pain and Pleasure are low-level occurrences in the Universe that do not happen to something but rather occur, freestanding.
- Pain and Pleasure are the result of properties of the Universe presently unknown.
- Pain and Pleasure can be evoked by some electromagnetic process.
- There is a Pain field and a Pleasure field (which may be the same).
- Base consciousness is literally a force of nature, occurring as a fundamental physical action.
- As with any force field, otherwise independent spots of consciousness locally add.

The rest of this book goes on to pose somewhat more specific possible details, mostly about the fundamental particles that make up consciousness. As it turns out these particles are not only conceptual, but literal.

The brain is a machine that, among other things, causes consciousness, as a means to the end of choosing better actions to promote the survival of the individual and therefore its genes. The only reason we are conscious is that it is a useful trick to propagate genes. It is a tactic, over which we obsess. The "goal" was never to think and feel and speak, but to survive and reproduce.

Conscious feeling provides the root of Value, which directs the cognitive machinery. Consciousness also eventually allows for so-called common consciousness, which aids cooperation.

Some philosophical works notwithstanding, this book confidently presumes that consciousness, first of all, is real. Furthermore, in developing our theory, we do not quibble over (the interesting question of) what "real" is, as this does not appear fruitful or necessary to the discussion. We do dismiss, as peripheral at most to our physical theory, the view that all reality flows from consciousness. Reality is discussed in Appendix I.

Current hard science embodies everything that is *not* emotional or feeling, as specifically denoted by our phrase 'the *physical* universe." Little is written, especially from a scientific perspective, that addresses what base consciousness *is*, only what it correlates with in brain region activity or in psychology.

Once we admit that consciousness is real in our universe, our theoretic formulation is driven by two basic steps:
1. We introduce consciousness particles, foremost because they need to be free-standing from any observer, else what is feeling *them*?
2. We form a force argument, by looking at how physics has developed.

2. Defining terms: Levels of Consciousness

The term "consciousness" refers to processes varying from the low and minimal to the extremely elaborate. For example, one possible layering of the complex thing we call "consciousness" is:

language-inclusive/organized thought

mind, cohesiveness

basic awareness

cognitive consciousness (brings in information processing)

linked mass of consciousness

physical consciousness

In this book we divide the many meanings of "consciousness" into four levels.

2.1. Human

While the math of fundamental physics is exact, daily thought and operating rules and their objects are fuzzy. But our language, including math (and certain standard practices with various kinds of diagrams), allows the precise recording and manipulation of facts and words and concepts that are themselves vaguely defined. While the things manipulated by language are (semantically) fuzzy, a language can be (grammatically) exact. This adds a new capability to animal minds. Originally evolved to communicate originally simple thoughts to others, language became useful and then vital as communication to self. Today, without language, one has only animal thought.

There are some very interesting things about language. One can describe anything, even things that change with time, using only words, without any pictures or digits. Language holds *meaning*; words mean things, they are the carriers of packaged meaning. Words can convey all experiences that the receiver has enough of the right experiences to understand. Language captures all concepts about thinking and feeling and doing and being.

Language organizes thought such that complex and long-term actions and plans can be formulated and remembered, as well as communicated outward and recorded. (On some other planet it may be that a second organizational means has evolved for such internal thought organization, such as even higher reliance on spatial manipulations). Sophisticated long-term memory has also evolved.

Nouns carry no sense of time. Verbs describe action, which requires time, and prepositions tie together objects, time, space, and the abstract.

2.2. High

High consciousness is the realm of primates and dolphins. Essentially it is everything that is needed for human-level consciousness except the precise and organizing capabilities of language. This large level is everything beyond the simple registering and identification of individual impingements upon an organism (low consciousness).

2.3. Low

This is what is commonly called sentience. It is what the simplest organism has that we are willing to call sentient. It is what happens in such a being. It is the beginning of Mind.

2.4. Fundamental

This is the dominant focus of this book. We address the question: "what are the most basic entities and/or processes in our universe that make consciousness as we know it possible?"

Everything above Fundamental we refer to as "Higher Consciousness" (that is, higher than fundamental).

A key characteristic of our total consciousness is that the *feeling* is merged into a unitary big feel- a merging that suggests that locality, or connectedness, or both, matters. (We suspect the feeling causes more neural stimulation and firing which causes more feeling (positive feedback).)

With *cognitive* consciousness we say, "I feel *that* such-and-so." It is always *about* something (as too is language, which is fundamentally cognitive activity).

Brain activity gives rise to what we experience as higher consciousness and the activity appears to be the separate activity of individual neurons stimulating each other in concert. This individuality of activity in the brain supports the notion of local fundamental dots of consciousness.

So, what is fundamental consciousness (which we also call "base consciousness")? Imagine making a computer program that simulates you so well that no one can tell it's not you. Each time you are supposed to feel, variables in the program will track your expected feelings and step in for them. This program is not your actual clone that feels. It is a simulation that just pretends that it actually feels (despite extant writings of some to the contrary). The part that is missing, real feeling, is the fundamental consciousness. All the mechanisms of higher level consciousness could, in principle, function on a different, simulated substrate. The difference is at the bottom.

PART I:
FUNDAMENTAL CONSCIOUSNESS

3. Kinds of Consciousness

In the Introduction we noted four categorical layers of "consciousness," building up increasingly sophisticated elaboration.

Let us now notice that there two basic *kinds* of conscious "experience." The first kind is pain and pleasure. Pain and pleasure don't have to be about anything- that is additional information that is often associated. One may feel simply euphoric and/or dysphoric "for no reason." This consciousness has negativity or positivity, and only that. This we call *emotional consciousness*.

Now we deal with the problem of qualia. You can "feel" both emotions *and neutral occurrences*. Here the word "feel" is overloaded to indicate two different things- awareness of an informational stimulus, and pain or pleasure.

There is dispassionate impact. This is caused by some distinction appearing that is large enough to cause it to be "noticed"- possibly very much noticed. But we "don't care" about this thing that has appeared, one way or the other, it just is what it is. This we call *cognitive consciousness*.

Awareness of a thing (cognitive consciousness) requires observation *of* a thing (even a color is a thing, among many other things that could have been observed). A thing, though, can simply be, or change. Feeling (emotional consciousness) is a thing, or a thing changing. It does not require observation to be.

Cognitive consciousness notices and observes ("those") things. Pain and pleasure consciousness, you live in. What does this clue mean, and what happens at lower levels?

It is known that different chunks of our brain are cognitively engaged from those that are emotionally engaged. Different brain networks have the value (emotional) vs unvalued (cognitive) consciousness. (While a low-affect person can still think, a person with zero emotion probably could not engage in productive thought- there would not be enough to guide it.)

We can't tell for certain whether at bottom there is such a thing as a neutral truly fundamental bit of consciousness (impact), or whether, like a complete atom or set of atoms, there is no resulting *net* emotional charge, but rather plenty of ripples at a smaller scale, making an energetic "patch of noise" that is neutral but not nothing.

Let us now lay down a few facts for whatever clues they may offer.

Mammals have well-developed emotion. They evolved for close-to-the-ground night foraging while the reptiles slept. Their olfactory regions elaborated greatly. Mammals developed the ability to learn (and later teach) more advanced things. Those that felt emotions that kept them long near their young, replicated more successfully. Social cooperation and competition flowered. Judgement as to what was important to remember or act on flowered, and largely replaced the perfunctory past of more primitive animals.

Then, proto-primates with the fully-developed emotions took to the niche of the trees and a phase of huge 3-D visuomotor development took place.

Now, as humans, it is difficult not to have at least a little of an emotional response to almost anything. An innocuous event may startle and cause dysphoric anxiety and stress and fear. You may notice a color to be drab or pretty, even if that is not relevant. You may notice a number to be an exact multiple of 100, and for some reason (neat order), you like those. Pleasure responses evolved for occurrences useful for thinking, such as the pleasures of understanding, discovery, novel adventure, order, and matching.

Where an action is largely "unconscious," like driving, there is no emotional response. As soon as it comes to the forefront you may feel the slight pain of exertion to focus on it or the joy of a result, or both at the same time.

Notice the use of the word "unconscious" in the last paragraph. Your brain is acting in full recognition of sophisticated and complex constructs while it drives, yet in ordinary cases, that has no emotional intensity. There we have pure cognitive consciousness, devoid of emotional response.

In general, it is nearly impossible to have something very impactful with no elicited emotional response at all. A great big cube suddenly projected on a screen right in front of you will have a little fear or delight associated with the "pow."

Change causes attention centers to activate, but you may handle the situation "unconsciously." "To really get your attention," which means to bring it strongly into your consciousness, someone may make the experience painful or delightful.

Would we find two kinds of *fundamental* consciousness were we to drill all the way down- a neutral kind and a good/bad experience? Possibly so. That would be but a detail to the theory we shall present. But more likely not, as we will see.

Besides the Occam's Razor argument, there is that the mind performs its other functions using ultimately only the extremely small set of characteristics of the universe that are available (and not even all of those), such as matter particles and electric charge. Scientific experience with nature tells us that something so vastly higher up as what we commonly call consciousness, is probably built out of a very few things at the bottom.

In which case, what we have is cognitive attention and emotional consciousness.

A reason it is so hard to completely not care in any way at all about something presented, is that everything is extensively presented to our emotional centers to determine its value, and rivulets of emotion overflow even if there is little net emotional reaction. Emotion will cause us to increase our attention but what we are conscious *of* is the motion that sent us to the situation. ("Of" is a reference to a thing. "Of" is cognitive, not emotional.) "Cognitive consciousness" is attentional.

What makes our human attention able to be so strong, is that our brain is made so that the centers that handle the new and current are linked in to large emotional centers.

At first what you are working on is very conscious because you *care*. You want the result and are imagining that pleasure, or you are concerned or a little tense about making mistakes. Your emotional centers have been activated and make a soup of individual little reactions (that in large part cancel each other out- more on that later).

One might claim that, at some level, pain and pleasure are not just the opposite of each other because it is possible to feel plenty of both at the same time. However, gross pain and pleasure don't happen at the same places in your brain. They happen in their own neighborhoods, so they don't locally add and cancel.

A concentration and synchronization are needed for a mob of fundamental pain or pleasure events to fuse and emerge- to mutually sum. Your most basic large-scale consciousness is the cloud of fundamental consciousness events in a locality. (From a quantum field standpoint, as you walk the cloud moves with you, rather as a magnetism travels with a moving magnet.)

It is also possible that a low-grade higher consciousness first emerged composed of largely disorganized and non-directional batches of fundamental consciousness bits, and mammals (evolutionarily) discovered how to concentrate it in one direction to get effective higher emotional consciousness. In that case also, the fundamental is the positive/negative consciousness and not the neutral.

Finally, let us suppose that there really is also, fundamentally, consciousness that is neutral. Emotional consciousness we expect to be easier to detect, localize and isolate. In any case, it is very likely that pursuing a theory of only emotional fundamental consciousness will cause the mystery of any non-emotional consciousness to quickly fall, since it must explain any supposed neutral consciousness in addition to the undeniable emotional consciousness.

It is likely that conscious "impact" is intrinsically emotional at bottom, and in proposing a specific theory, this book proceeds on that belief on that detail.

To review, there is an important difference between the cognitive and emotional kinds of consciousness. With cognitive consciousness you experience an awareness *of* something. It is disinterested, almost arm's length. *Pleasure or pain you are not observing from the outside like that, they are something you live In.* Consciousness does not appear to be of only one kind, at least not just the cognitive kind, which requires both the raw capability to be conscious plus the thing to be aware of. *Cognitive consciousness requires an observer; emotional consciousness does not, and is therefore a solution to the observer problem.*

While cognitive consciousness is always about something, pain and pleasure are not. You may say "I hurt," or" I feel good," but you don't say "I noticed," you say, "I noticed X." *Pleasure is not about something, rather it is merely caused by something.* Neutral consciousness ("impact") *might* exist, and at higher levels at least it requires an observer. But we know that emotional consciousness *must* exist, and requires no observer.

While object perception can be complex and require appraisal, even a trivial perception such as a nonpainful pinpoint prick of the skin feels like one is observing it from some distance apart from it. Our cognitive consciousness originated about physical senses of things outside of the body. Emotional consciousness is about our internal value system reacting.

It is unlikely that for every basic thing we, or another planetary race, could sense, there is also a unique and fundamental action, in physics, of awareness of it. Therefore, consciousness of color, for example, is not fundamental; there is something smaller.

The bedrock stuff for consciousness is the feeling we are in. *These are the feelings which themselves feel, freestanding.* In us, thousands or millions do so in concert, located as it happens in the space of our head, interconnected by cognitive neural links and feeling neural links, and immersed together in a field as we shall see, resulting in a group-feel mob.

Abstract things that are felt to be pleasurable, such as color combinations or music, are clearly artifacts of abilities evolved for important real things. Today we refine our means of activating these pleasure sources and make candy and paintings and songs and so on.

The above discussion makes us conclude

1) fundamental consciousness is positive/negative
2) fundamental consciousness feels on its own when it is made to happen and is not observing, or doing some composite process
3) One or two fundamental units of this consciousness happening don't amount to much.
4) There is some kind of communication between the units, that cause the activation based on the activation of neighbors (this is the field, to be described later).

Emotional feeling exists when it does, on its own. It is not experienced by something else, it just is. It is a fundamental phenomenon of nature. Pain and pleasure have opposite signs. That doesn't necessarily mean they are the very same force, but the simplest model is of one force.

It is interesting to note that at your earliest, you have no thought, but you do have feelings, to your deep beginning.

The existence of largely neutral impact experiences subtly suggests that (net!) neutrality is the more normal and natural condition, and that some things asymmetrically concentrate negativity or positivity in a place.

4. Emotional Consciousness is the Foundation

In common discourse, there are two meanings of the word "feeling." Feeling can be an act of cognitive consciousness (you can feel internal brain/mind things or an external touch of neutral character), or one can feel good or bad. There are not one but two heretofore-not-understood phenomena.

There is the quizzical case of neutral feeling (impact), a fascinating example being the response to seeing a colorless and shapeless splotch. With the lack of emotional content, we would call the splotch "plain" or "ordinary." Impact does not have to be emotional, but it "registers"- deflects/moves something in the fabric of feeling. When you are conscious that you see a truck- the *recognition* is complex and associative, the *impact* is not.

Let us now explore impact more deeply- what its nature is. In the case of something indifferently noticed, the two possibilities are that either plus and minus cancel out- which leaves the problem of *strongly* noticing something you don't emotionally care about- or the existence of fundamental consciousness devoid of emotional feeling (pure "impact").

Thinking on this suggests one model where absolute displacement or absolute speed of emotion adds for a related phenomenon of impact, a bit like magnetism arising from moving electric charge but, like gravity, without the polarity. Emotional "froth" equals cognitive impact.

And then there is this: Suppose one of the two kinds of consciousness is like charge, and one is a result of that charge moving (like electricity and magnetism). In our universe there is no absolute direction, no plus vs minus, so sign must be ascribed to the charge, which in this case is the emotional consciousness. Thus, in this model, again, impact is the consequential one, and emotional consciousness is the fundamental one. As with electric charge, emotional charge is signed.

At the higher consciousness levels, impact is always about something, in the cognitive neural network. That itself does not prove that there is no fundamental impact. This impact would be the tiniest imaginable wisp of consciousness, having neither value nor reference to anything, the neutrino of experience.

Can you have impact without an observer, without reference to anything? This is very subtle. The answer in theory is yes, one can have a sense of impact (a feeling of "something's up"), before the association to the object causing the impact, even though impact will always be caused by a thing, and that thing that is likely to enter higher cognitive consciousness.

There are presently no known precedents of triplets of particles in nature that feature signed and unsigned particles that are otherwise identical. But among the nonidentical, we have the weak force bosons, with the two W's the same and the Z different but for the same force. Consciousness could be weak-like. (This is interesting because of the connection of consciousness in the brain to the electromagnetic force which is part of the electro-weak interaction. A likely unification to be discovered anyway is the electroweakjoy force.)

Amongst our close family of proto-theories, we allow Occam's Razor to cut this possibility away for our continued analysis and synthesis, for if we first posit consciousness with zero emotional charge we must add negative and positive consciousness to get to all extant consciousness.

Awareness made from net neutral impact, in principle, may have been discovered by evolution first, like humans discovering a spark gap transmitter with no specialization of frequency or direction. But a feeling of naked awareness has neither utility nor reason to exist or be discovered or cultivated or formed. Indeed, neocortex did not explode until developed emotional cortex was present.

Neutral consciousness provides little evolutionary value in the absence of emotional consciousness with which to make decisions, so emotional consciousness, at least weakly organized, was probably discovered and harnessed first by evolution, during the process of evolving High consciousness.

You can't get from an *is* to an *ought*. Pain and pleasure appear to be fundamental. They are free-floating. Pleasure and pain are each a noun without a requisite *of*. *They* must exist.

This solves the observer mystery- things bounce off the brains, causing free-standing feelings.

With no emotional consciousness at all, we might not feel anything. Feeling a big red block may be a sort of side effect, as notification is always passed to the emotional centers for judgement, which then in this case have a weak or chaotic reaction. Pure cognitive acts could be done by animals without consciousness, as bugs.

Low-level or scattered excitation of the same neural system that provides emotional feeling may constitute a resultant cognitive feeling.

What makes the emotional centers of our brain is that they concentrate feeling that would otherwise be noise (impact), by making each base event happen at the same time or in the same direction or at the same frequency. Impact then is like white noise. Emotional feeling is less random.

Next, we check our thinking with this approach: Imagine you are waking up from anesthesia after a procedure. The probable sequence as you reboot (become fully awake) is:

1 first glimmers of vague disordered consciousness

2 feeling

3 cognitively aware of your happening feelings

4 conscious of your surroundings

5 organized thought with language

Basic feeling, to human experience, is awareness, registration. But, further, it is awareness *of* something. There is probably an even baser consciousness which is just feeling, not feeling about. The characteristics of human-level consciousness can trip up our analysis, because it is always engaged with cognitive circuits so cognitive awareness always occurs. That a feeling human mind will also be aware, does not mean awareness is fundamentally required in order to feel.

Consciousness probably serves a function. Our totality of it would be both written to and read from. It is a mechanism providing values for associations and decisions. It provides brief memory and communication.

It was not necessary specifically to feel, but we do. Feeling exists in our universe, and evolution has exploited that phenomenon.

To a human, what are the most basic feelings of all? Pain, pleasure, and impact. We have seen that the simplest answer, and therefore the answer likely to be close to correct, is that there is only one kind of fundamental consciousness, and if there is only one kind, it is emotional, not cognitive. Spots of emotional consciousness can add to approximately zero, whereas zero can never add to positive or negative. Fundamental consciousness can be, and can only be, euphoric or dysphoric.

5. Clues from Brains

Normally, a nerve signal, and only that, can cause us pain or pleasure. Therefore, pain and pleasure are neural, that is, something that happens in the nervous system causes pain and pleasure.

Compared to their predecessors, mammals are hyperemotional "drama queens." There is something different about emotional neural tissue; it is relatively more present in higher animals.

One path for examination would be scents because there is less neural complexity involved than with the other senses, both in early processing and in quantity of neurons to pass through to reach emotional centers. Scents can be very emotionally evocative (and very quickly). Cognitive consciousness is thalamically-related (and about the physical senses), with emotional consciousness being limbic (and an outgrowth from olfaction).

The construction of our brains is such that probably all the micro-spots feeling pain or pleasure are near others (or close together inside just a few different macroscopic places), and possibly activating synchronously or near together in time as well. It is possible that one or both of these traits matter to bring the unified feeling. In fact, in this book we expect this is so.

The most striking and useful fact is that our physical brains, which otherwise operate with physical processes that we understand, create- or connect to- consciousness.

Ample evidence suggests that neurons are the elements producing consciousness. What mostly happens in our neural networks? Various operations on a molecular scale, most of which look just like those of other cells. Those other cells may even be weakly conscious, but naturally we should look at what makes neurons fundamentally different, and the most striking and defining difference is that they produce and transfer action potentials. Furthermore, interfering with the membranes carrying these action potentials, such as with widely differing anesthetic drugs, removes consciousness.

All cell membranes have ion channels, but voltage-gated ion channels are characteristic of neurons (and of muscle cells, upon which neurons impinge).

And what is happening that makes up an action potential? Movement of electrical charge in a 3-D pattern that is very unusual in nature, on a small dimensional scale, along a constant-diameter axon. There is a locally distinct net concentration of charge moving along a linear path for a significant length. At the same traveling location there is, orthogonal to the that movement, a circular ring of charge movement across the walls of the small-radius axon. In many parts of the brain these axons are numerous and dense, and in many parts, they run in parallel in the same direction.

What is going on in the brain is movement of matter and ions, and consciousness results. There is nothing about the movement of mass that looks importantly unusual. Gravity does not have any obvious significance in this motion, nor does the strong nuclear force. The electromagnetic force does appear to have a relationship to consciousness, as all the unusual neural firing and flow is of ion charge.

As the charge carriers are nontrivial ionized atoms, the charge is primarily determined by the electron situation. But it is interesting that the current is of ions, which include nuclei, and not of electrons. These ions are elemental; the charge is not carried by molecules.

Another clue from the brain is that it is a molecular machine, operating at scales that can effect local quantum state changes.

We surmise that either the electromagnetic actions of the spike directly cause the conscious event, or that the spike causes molecular events that in turn generate the conscious event.

6. Who is Doing the Feeling?

This age-old question we answer thus: the feeling is doing the feeling. Feeling does not happen *to* something, feeling just happens. Feeling, for a moment, *is*. Feeling can be at a place that is contained by something, and happening as a thousand simultaneous points of light. It *is* experience, as surely as a photon is light. In fact, it is the only thing that is experience, in the conscious sense! In other words, Leibniz was basically right (monads).

In your head are millions of points of feeling affecting each other in various ways (more on that later). This swell of activity is what "you" feel. Along with your unconscious skills and your physical matter, it is what "you" *are*. I feel therefore I am.

7. Real, Natural Phenomena Are Described by Physics and Math

There are two meanings of "real," two meanings of "exist." One denotes what we call physical presence. One refers to rules or form, typically a codification of what can exist in the first sense of the word. (Either kind of item can also be hypothecated but not be real.) And there is an issue of "absolute" vs. "relative." These matters are discussed in Appendix I.

What we call Reality, is what exceeds just the math, what has presence, whereas in math to exist only means to be possible and there is no distinction between the two. Math is the patterns of the real, observed from the real. There is no mystery or need to philosophize why mathematics, created by humans to describe physical processes, works, because the brain was developed in this natural world with its rules. (In places the brain has simplified from what is strictly true- such as with hard-edged objects or absolute causality, but while this does not work for correct truth, it does work for effectiveness.)

Physics, which uses math and largely is math, acknowledges, assumes, and is about the "hard" definition of "real"- the physically present.

Exempting, then, the math definition of "real," what is Real? Matter, radiation, and feeling. Starting with a few brute facts, these real things can be shown to be a deductive consequence of the rules we have discovered, except feeling. The generative metaphysics is math (except the math doesn't declare how much stuff there is, with 0 being only one possible value). Only anthropically is feeling required to exist, we have no rule of physics predicting feeling (yet). This suggests there should be some math predicting feeling. But how do we know what that math would look like? With feeling we are inside it, we are it. What the math needs to predict is the boson and the field that corresponds to the action we call feeling. The math does not have to predict why feeling feels, it only must predict when and how it happens (our math does not predict what it is like to be a traveling photon, after all). Some other property of the universe that would similarly provide value in decision making and memory, would (and perhaps somewhere does), serve, and might even bear functional similarities to feeling while not being feeling.

Physics describes our universe. Our universe consists entirely of particles interacting in spacetime. This behavior is described by mathematical laws. The different interactions we call forces.

The interactions vary in size and intensity, which we quantify with the mathematically derived concept of "energy." The universe has a nearly equally-distributed energy content, which drives the possible interactions. The interactions have ripple effects to other interactions, and the total calculated energy remains the same.

We model interactions (forces) with something called a "field." In classical pre-quantum physics, a particle with a certain characteristic, say electric charge, creates by virtue of that charge a force field around it (which trails off in intensity with distance). Another particle in the first particle's field which also has charge, will experience a calculable force upon it, and this push or pull will cause the particle to move if it is free. (The second particle will also create its own field, summing with the first and affecting the first particle).

To the degree the particle(s) are not free to move, "potential energy" will be present and is typically said to be "stored in" the field.

While at our macroscopic level, things effectively do or do not happen exactly because of specific reasons, this is not the fundamental way our universe works. Simulations suggest that a completely specified and deterministic universe would tend to ossify and crystallize into more primitive structure than the one we find teeming with life and intelligence. But an anything-goes universe with no order would remain a grey soup of background noise from which nothing interesting would emerge, either. (Universes like these likely exist "somewhere" else.)

Our universe *constrains* what may happen, and how everything must balance out on average, but does not micromanage what at any given moment *must* happen. The laws of the universe describe precisely what it is *possible* for the particles to do (interact).

Humans, at their scale, are often uncomfortable with this notion, and because of the nature of their brains and minds tend to think everything that happens must be justified. There is no *universal* preference for this, however, or any deep reason there should be. Apparent strict causality is an emergent illusion.

This discovery about the universe, applied to updating our model of the universe, is called quantum physics. It explains everything physical except gravity, and the gravity math is being very actively worked on.

In quantum physics, the mathematical construct that is the classical field is replaced by a quantum field. A quantum field is a physical entity, with energy content and structure of its own.

An equivalent representation of this field is as a field particle, also known as a "boson." A boson is the "force" carrier between two other interacting particles. Each force has its own kind of charge and its own field and boson(s).

In the current next step in the development of our physics to date, known as M-theory (which is still incomplete and unproven), gravity is brought in to the math, encompassing "all" the forces. In this theory there is a slight shift in that each of the boson types and each matter particle, too, is seen as a different vibration at a different frequency in the same single field. This effectively means that "empty" space contains everywhere this single field, which contains energy.

Consciousness is a real phenomenon, and for any Theory of Everything to really be a Theory of *Everything* it must address consciousness.

Math works because our minds evolved in this universe. Math is a capturing of the laws of nature into enough neural circuitry, by evolution, to allow us to formulate it. Particle and force have been shown to a*lways* be the explanation. Consciousness is a physical thing- a physical thing describable by physics, to be described as deeply as we "understand" what light and matter "are."

8. The Consciousness Field and The Senton

8.1.　　Basic Argument

All natural happenings are explained by forces and their fields. We know that consciousness happens, and it happens to us, in our natural world. While it is mysterious (as have been many phenomena for a time), there is no logical reason to assume it is somehow not natural. Quite the contrary, for modern man, that would be a bizarre (and popular) leap to make. Either all things real are describable by physics, or just not this one (consciousness), which is unlikely.

Other basic actions are known to exist, such as gravity and electrostatic attraction. All of them are particle interactions, or radiation when a boson has not reached a target. All actions so far understood are movements of particles in space.

It has long been suspected that consciousness is a natural and basic component of the universe. Aristotle viewed it as one form of energy, and Leibniz (presciently) had his monads.

We *know* there is a connection to existing physics because of the brain. The brain evokes consciousness, and the brain is made from matter. The brain is made only of particles. It either contains or interacts with what is needed for consciousness. We know that consciousness is something caused, because the brain causes it.

Fundamentally, all pain is the same- dysphoric. All pleasure is the same- euphoric.

The force that we experience as emotional feeling shows itself to be very different, as we can build an entire physical theory of (almost!) everything without mentioning it. The *only* thing that tells us that there is feeling, is that we "internally" experience it.

It is because consciousness is so different from the rest of the phenomena, and that we can construct an entire model of the material universe without needing to include it, that it is first suggested to our minds that this a "new" force. (It is also possible that it is another effect of an existing one, in the way that electricity begets magnetism.

Let us consider a similar mystery, recently solved, that of particle rest mass. The simplest explanatory mechanism adds another quantum field, that permeates all space. This "Higgs" field would cause spontaneous symmetry breaking during interactions. The field would have its own corresponding boson (the Higgs boson). This particle was searched for, and found, confirming not only the mass theory but the strength of the existing physical model of the universe, with which it is compatible.

Another case of a new field is the hypothetical field responsible for cosmic inflation- the inflaton field (like the Higgs, this would be a scalar field, possibly a modified version of the Higgs field). Yet another case is dark energy, which is speculated to be contained in another scalar field. A field theory to explain gravity calls for the existence of gravitational waves; these were detected in 2015.

It is possible that consciousness is not only action in a new field, but that the field is one of these same fields.

There are two *known constituents* of the universe that are not understood, missing explanation. These are dark matter and dark energy. The question naturally arises whether consciousness could be related to either of these.

Dark matter, while interacting gravitationally, does not interact with electromagnetic radiation (light). Since electrical flows in the brain appear to produce consciousness, this suggests that dark matter is not closely related to consciousness.

Most of the energy in the universe is vacuum energy of space, also known as dark energy. It is conceivable that this is consciousness. Dark energy may be linked in an unknown way to the "normal" part of our universe, much as normal electromagnetic behavior appears to be linked in an unknown way to consciousness events. Dark energy is particularly intriguing because though we know it exists, it doesn't seem to "do anything." What is its full "purpose", its role, in tying our world together?

Let us now make two observations. First, all physical phenomena turn out to be describable by refinements of our standard mathematical models. To date there has been an effort to find a so-called Theory of Everything. No theory that leaves out consciousness is a theory of everything.

Second, fundamental consciousness happens on its own as an essential freestanding phenomenon- that is, it does not at bottom happen to an experiencer (nor could it), it happens itself. Pain is not experienced by something, at its fundamental level, it simply occurs. (And "I am in pain" is a far higher-level concept, as is "I".)

In every case so far where a very different fundamental phenomenon occurs in the Universe, a particle characteristic quality has been identified that correlates to this. In most cases (all but spin), an associated force field and attendant force carrier particles are associated with this quality. Particles carrying multiple kinds of charges can cause cross-force interactions: a particle that has both mass and charge can be moved gravitationally; its change in position causes change in its effect upon other charged particles around it.

There are four known forces, plus the omnipresent, scalar Higgs. If we examine the brain, we see nothing happening with the strong force. Rather, the entire load of what it does, of that about which we already know, is being carried by the electromagnetic force, and by matter, in its extended space and time. All present experiments suggest that electrical activity (be it at the level of electrons, or the molecules that are charged, shaped, and interact via electrons), "cause" consciousness, though we have not been able to trace this process down to a fundamental act.

This means that probably consciousness arises either 1) quite directly consequent to a certain electrical event, such as do magnetic fields that emanate from moving electrons, or 2) by causing an action with a heretofore unknown force or particle quality with which the electrically charged particles can somehow interact. Alternatively, or in addition, matter containing sentient charge may be harvested from the environment and stored in the brain.

Now even if it is the first case- say a precise oscillation frequency involving enough electrical charges- whatever is then evoked within the electromagnetosentient force is not yet understood, and is separate at least as much as is electricity from magnetism. What we seek therefore is the sentience field or the sentience particle quality.

Next, notice that even naked logic, naïve of physics, suggests that there needs to be not just a particle quality- a charge- but a corresponding field, because the individual actions need to sum together, which is something a field does for us. This particle quality is "emotional charge," and like every other known force field, corresponds to at least one boson, which we call the senton (from "sentience," the capacity to feel).

The fundamental consciousness, which is emotional (positive or negative feeling rather than neutral), involves an emotional charge. Following the lead of truth and beauty (which are quarks), we call this joy charge.

Consciousness is a fundamental phenomenon. It is not magical, but it is fundamental. Since one can't get from an *is* to an *ought*, one must add emotional consciousness to the existing physical description of the universe and to any neutral, cognitive consciousness.

As a fundamental phenomenon it neither has nor needs a Why it happens. It is a base feature of our universe.

There is definitely a new force or sub-force, and a new field (joy). These correspond to a boson (conceivably the photon), which may be a new boson (the senton), and a charge (possibly electric charge), which may be a new charge (joy charge). We speak of and assume sentons and joy charge, and characterize them, until such time as it may be proven that one or the other is identical to a known item.

8.2. Deeper Exploration Notes I

Fundamental pain and pleasure did not *evolve,* in order for us to survive (although our anthropic universe featuring them, allowed us to), because there is no way to get from cold-physical to feeling. We do not just know that we feel or act like we feel or are hardwired to do certain things as if we feel- that is what simulated consciousness does. We *actually feel*. Like and dislike (pleasure and pain) are fundamental qualities of our universe.

The interaction with the consciousness force is probably either with electromagnetism or with changes in quantum states throwing off or receiving energy.

It so happens that the boson corresponding to both electricity and magnetism is the photon. It is possible that electric charge in, say, accelerating, could emit oscillation of the sentience field, much as how certain motions by the electron can emit photons, and in this case the photon might also be the carrier of sentience (perhaps a third axis of light). This seems somewhat less likely than other possible theories because there appears to be something very specific that needs to happen to produce consciousness; we have bumped into no evidence of very widespread appearance despite widespread manipulation and actions of electrons and photons. More likely, would be that a photon absorption/emission causes a state change in some other particle or system of particles (which is often the case with the understood phenomena). (As with other bosons, the photon may or may not carry the sentience charge itself, it may just convey energy to something else that does and change its quantum state.)

(Technically it is possible that we have already created true artificial consciousness with our electronic devices without knowing it (raw consciousness, not a mind).)

Another clue for us on the exact nature of consciousness, is that it would have to be discoverable by evolution, starting from whatever had evolved already.

Consciousness bosons (sentons) must be compatible with the (possibly extended) Standard Model. The consciousness mechanism should be quantitatively describable, and in principle, measurable.

In principle, three possibilities allowing neutral as well as positive and negative consciousness, are:

- Multiple new forces
- A new force that features multiple kinds of bosons
- Effective co-cancellation of positive and negative (there can still be simultaneous surviving positive and negative clouds due to spatially separate brain regions)

The last is the simplest.

Regardless of their exact nature or location, we expect free-standing spots of fundamental consciousness- free-standing in the sense that they are themselves the feeling and nothing else needs to feel them. They are completely disembodied. These separate bits influence nearby others to fuse in the moment into a unified construction.

The simplest solutions are that either the joy force is another effect of the known electroweak-ness, or that the known matter particles carry the new joy charge, either electrons or quarks. Interestingly, neural-specific flows are of ions, not electrons or protons, so there is no clue there as to which fermions might be involved. And interestingly, both electrons and quarks carry electric charge. It is also possible that joy charge is carried not by known fermions but only by known bosons, which would explain difficulty in observing the direct link between neural behaviors and consciousness.

Once the theory is comprehensive enough, feeling will be one more resonance in the common M-theory-style field. This field is the new ether, the medium for feelings as well as things, and the interactions between them.

8.3. Deeper Exploration Notes II

Pain (and pleasure) simply is. It is not made of anything else. What is physically happening when this happens? Is pain the passive existence of something, or active- a particle, or an action?

Pain appears to be an action, a happening, or at least an unresolved potential difference- at the very least it is probably dependent on position or change of position. The existence of a pain (consciousness) charge notwithstanding, pain needs to *happen*. Our belief that this is so will likely lead our search to the truth.

For now, we don't need to know the exact sites and action of brain cells involved, we just need to know the process or the small number of processes.

Pain is an action. What then, is the action? It is

1. Motion or
2. Decay or fusion or

3. Change in a quantum state or
4. Change in a particle property

That is all there is, and there is no cause to suspect something entirely different. Everything else in the Universe works this way.

The simplest acts are the ones we should look to first. There is nothing to suggest that one particular frequency of radiation causes or is pain while another is not. So, it should not be at its bottom a particular frequency of a photon or other particle but something new and very simple and similar to what is already known, such as, say, the creation of a certain particle.

As with frequencies, there is also nothing special about arbitrary already-known kinds of changes in quantum state.

Pain is wholly different, and very isolated in interaction, in terms of physics (that we know). It should correspond directly to a particle or quantum quality or behavior that is not already known to us.

Perhaps certain nutrients contain this other charge, and they collect in the nervous system.

Pain we expect to be a transition of a fundamental thing. This suggests a new quality, like electric charge, and a new field. (One quality- spin- does not have one- particles can have spin but there is no spin boson or field. Of course, that, too, was our early view of mass. There may be more to know about spin.)

Consciousness is a process. A human feeling must take at least a little time in which to occur. A human feeling is something that happens, and that can happen with no cognitive context at first (such as euphoria or dysphoria). Therefore, a fundamental consciousness event must occur involving at least one state change. Such low-level events in physics include decays, spin flips, electrons changing orbit, and emissions of photons. Rather than a "point" event it could be a continuing one, like photon or fermion radiation, for example. At the bottom there are needed particles and forces- either the ones we already know, or new ones.

It is possible that the Standard Model does explain everything already, in that one of the operations known to occur corresponds directly to conscious feeling. But there is nothing in that theory that causes one to predict the existence of conscious feeling.

It will be interesting in years hence to look for a conservation law here (which could even prove happiness to be a fundamentally zero-sum game).

Consciousness could be a second-order effect, like magnetism via charge motion. Particle decays or creations could possibly be the acts that correspond to occurrences of fundamental consciousness. Fermions can emit and absorb bosons, such as Z particles. Does feeling happen when particle spin flips?

Is there emotional spin, which can be in one of two directions per particle, and aligned among particles (yielding an effective emotional magnetism) or not. When they are distended in an E-M or other affecting field and flip back, they could emit a senton or photon.

So far there is no evidence in favor of absorbing a particle or particle decay or particle creation or an elaborate mechanism. The simplest and most likely case for consciousness is of bosons and charges. Many known phenomena such as potential difference and electromagnetic resonance emerge from free-standing charges of individual particles.

8.4. Conjectures and Speculations

What would be the other properties of a feeling-charged particle? They apparently do not interact with much we know, or not intensely, for if they did we would have realized their existence sooner. Since we haven't discovered such a particle yet, it either

- Is one we know but did not realize causes feeling
- Is too heavy to have seen yet
- Operates at relatively short range
- Interacts weakly with the objects we know
- Causes changes of short duration
- Is very short-lived, or
- Is "hiding" in the seven small dimensions

Note that we have holes in a periodic table of particles. And M-theory leaves room for a lot more new particles.

Where might we find these elusive consciousness particles?

- At extremely high energies? Then they would be too short-range to add together well in a field.
- In the curled-up 7 dimensions? Then there needs to be an interaction that reaches into "our" dimensions.
- In any case, in the first tiny fraction of a second of the life of the Universe.

Pain appears to be a process/action, like a flow or a flip. Both states could be non-painful and yet the flip be painful.

For a particle that can have pain charge (joy charge of -j), its motions then might cause pain.

Perhaps state changes can emit sentons of energy, much as they do photons.

Neurotransmitter release is for some reason probabilistic; perhaps in that machinery sentons are sometimes emitted and sometimes not.

Where sentons are all released in near proximity in space and time, pain or pleasure could occur significantly. These operations apparently are in high quantity and simultaneous and in close proximity in the pain and pleasure feeling areas of the brain.

Perhaps the only thing sentons do to their interaction participants is change energy levels.

A neural spike has a forward flow and orthogonal, circularly-arranged flows. Perhaps one carries impact intensity and the other is emotional value.

Such organized charge (and magnetic) flows are unusual in Nature, appearing only in neurons. Perhaps flowing charge produces feeling, just as it does magnetism (magnetic force): a moving charge creates a feeling field. Light may be a carrier of feeling also.

It is also possible that the reverse of what we'd first expect is true- that clumsy perturbation of the joy field produces strong emotional feelings, and more evolved stimulus produces emotionally quiet impact.

Electricity and resulting magnetism are at right angles. In 3-D that leaves one more right angle available. In ordinary space, there is that third dimension of freedom in a traveling photon for a consciousness wave (namely, the direction of travel). The new right angle may instead be in one of the seven "other" dimensions.

A flip in dipole moment may cause consciousness.

Physical direction of spike flow might matter- the sameness of direction. This suggests what would be positive or negative emotion, in the opposite flow, cancels. That also suggests two different charge values (or fields) rather than a single one moving in one direction versus another, which would have to be true. A negative charge moving in *any* direction causes negative feeling. Other negative charges in opposing directions cancel it out. A positive charge in any direction causes positive; same for canceling. Thus positive vs. negative feeling would be based on which charge (such those as in septal area vs. those in pain centers), and sameness of basic direction.

Frothing (random joy field activity) may account for neutral consciousness (impact)- net neutral but noisy.

Let us say that the electromagnetic force is actually the electromagnetosentient force. Then there might be conversions/effects that are nonobvious in nature as with electricity and magnetism. The carrier would be either photons or another boson. Perhaps EM has a second boson. Consciousness does seem to be related to the electromagnetic force and it seems to be transitory- on the order of milliseconds- similar to action potentials.

For the joy interaction, there will be a Quantum Sensio-Dynamics to be discovered that describes it.

Joy may already have been exploited in a smaller way such as by reptiles, and mammalian evolution figured out how to organize and concentrate it like a laser. Now it is large and unified instead of local, weak, and semi-random.

Other physical events may involve sentons and be conscious-like in some way, without our knowledge. If *all* currents of charges caused emotion, we wouldn't know.

There will be consciousness bosons in other parts of the universe. What makes us is that we are each "inside" the consciousness interaction (other creatures in the universe may be inside other interactions, instead).

While charge flows made of electrons have no nuclear forces applicable to them, ions do. (Also, unlike classical electrons, ions have size.)

Extending from quantum physics, joy is a force field. If joy instead more resembles gravity, it is a distortion of other dimension(s).

Magnetic force depends on a particle's velocity. It is possible that a force could depend on a particle's acceleration. Joy could be related to acceleration, e.g., change in flow, instead of velocity or position.

Neurons are talking to each other, some neurons are conscious (create joy), some are probably separate and isolated and distant, some may not be very conscious.

The substantial and single "you" emotional feeling means the point feelings must interact. There must be effect locally between each tiny event to make the collective feeling, so fusing into one corpus. That demands the field, which will supply the additive and fusing nature.

Bits of plus and minus will nullify *if* they are all mixed by their proximity in space- this is one reason to suspect locality like other *fields,* and inverse-square like the EM force, which has nontrivial reach. It is also possible that it matters how long the perturbation lasts to be effectively emotional or not.

Some physical interactions might feature a decay to both a +j and a -j -carrying particle, resulting in a net 0 feel (and conserving joy). (They would still each have their momenta).

At the macro level, neutral impact is probably not just a lot of pluses and minuses, because the brain tends to AGC (automatic gain control- enhancing strongest signal and reducing the rest- a sort of mutual exclusivity). But at particle-level, it still is.

The joy field extending over a distance conflicts with living *entirely* in the small dimensions, unless there is an orthogonal consequent action, as with electricity and magnetism. EM having another orthogonal effect that takes place in the small dimensions would "work."

Sentons may themselves carry consciousness charge, as gluons carry color charge.

Consciousness is a perturbation or resonance in the quantum consciousness field. In some way there is an interaction between known fermions or bosons and the consciousness field. That means either the consciousness bosons also carry one of the known particle properties but have not been discovered, or some of the existing particles carry the consciousness charge, or consciousness bosons are involved in quantum energy level changes of systems. Given that particles can have multiple properties, there might not even be a new boson, just a new field.

8.5. Conclusions

Each feeling has

1. an intensity

2. a sign: dysphoric or euphoric

A new force mediates the consciousness interaction. It acts on two particles with consciousness charge, and/or may be emitted or absorbed by a quantum system changing energy levels. On a molecular scale, neurons may have these systems.

Something specific that is electromagnetic (electromagnetic being special because most of the brain's matter motions are of things with net charge, too, and other nonfeeling things move all the time), or some specific change that is molecular, causes a pain or pleasure event. A pain or pleasure event involves something outside of the known interactions. Barring some very special frequency or configuration or configuration change being the only at-bottom explanation, which seems unlikely, this requires at the least a new force and field. Associated, is emotional charge. In the positive spirit of the already discovered particles Truth and Beauty (and in deliberate conflict with the Russian Anthropic Principle), we call these "joy," "joy field," and "joy charge." All fundamental particle fundamental properties are restricted to specific values, so the same is probably true for this one. We call this probable fundamental charge constant j.

Sentons are hereby predicted to exist, to explain the pain and pleasure phenomena, and they interact with particles possessing joy charge of $+j$ or $-j$ (or multiples thereof). The senton is the joy charge boson; joy charge is the fundamental consciousness charge.

Rather like macroscopic action requiring cause emerging from probabilities, sentient beings experiencing emerges from fundamental consciousness events.

It has been said by some that acts of consciousness create Reality. Probably, it is rather that acts of Reality create consciousness.

In one way or another consciousness will turn out to be related to electric charge at a very low level, if only by what particles carry what combination of charges.

From the standpoint of established physics (the Standard Model), certain particles interact with joy fields. From the standpoint of M-theory, certain particles participate in joy interactions with the always extant common field. The difference has no effect on our arguments in this book. More often, this book speaks in terms of the joy field rather than the hypothesized common field.

8.6. Further Conjectures and Speculations

Given a joy force field, there is a force constant. We have gravity between each other, and perhaps as well, joy force. In the case of gravity, we don't feel it between us because the force constant is low. In the case of electric charge, we also don't feel the force because the positive and negative almost exactly cancel out.

We might expect an inverse square law for base consciousness for three reasons

- sameness and simplicity
- apparently, the joy-charged particles interact with EM charge, which follows the inverse square law
- range is needed for addition on the scale of neural tissue

Notice that you don't actually know that inanimate objects don't feel. In fact, if your brain is cut in half, each half doesn't even know that the other feels.

An interaction occurs when bosons are exchanged. Radiation occurs when bosons are transmitted into space. This predicts joy radiation, transmittable and receivable, and probably travelling quickly. Is it subject to special relativity?

Now let us examine known elementary particle fundamental qualities:

Spin: For a charged particle this results in a magnetic field. Photons (which are not charged) also have spin, which determines their polarization. There is no separate field or boson known for spin.

Mass: gravitation field, graviton, Higgs field, Higgs boson, neither boson has spin 1

Color charge: field and bosons

Electric charge: 2 fields and 1 boson (no magnetic monopole in proven theories)

Weak isospin (weak charge): +/- ½ it was discovered that all elementary matter particles have this (including dark matter).

Interesting is that a) the weak force is closely related to the electromagnetic force, b) the electromagnetic force is fundamental to the neural spikes, and c) the weak force connects all matter, including universal constituents not yet well-understood. It is possible the electroweak force (which by far is already responsible for the vast majority of actions we observe) includes further interaction types of which we are not yet aware. As usual, one class of experimental approaches to discover these, would be to look for missing or extra energy in interactions within neurons or brains.

Spin can be changed in its orientation in space by magnetic fields or photons.

Photons change spin of net charged particles. Weak force Z particles change particles' states, too.

Precise resonant frequencies flip spins of existing particles- energy coming in. Same could happen for a "pain spin." Then the specific construction and behavior differences of different neurons becomes interesting.

There are molecules in the eye (which is an extension of the cerebral cortex) that react to bosons of light energy (photons). There could also be in the brain molecules that react to bosons of joy energy.

We can simplify by just looking at pain, presuming pleasure would be by a similar process.

Pain does not take up space and neither do bosons or force fields. They are just as invisible as pain is.

Pain "just happens" in the same way that the pull you feel between magnets just happens, and magnetic fields just happen when electrons move or spin.

It could be that still, moving, or accelerating electrons also make a sentience (joy) field, but it is usually distributed and randomly oriented, so unnoticed.

As much as we know then is: physical quality of particle (including location), changing. Some electromagnetic operation makes this happen.

Emotional charge may be a property of all matter, or of some matter. In certain cases, as with magnetic material, there is a lining up of movement or other change in the same direction.

If in a neuron a certain resonance is achieved, then what? Probably not just the resonance itself, nothing special about that, it needs to resonate with something else. That something else is, at a minimum, a quality of a particle (or quantum system like a molecule).

A photon or other electroweak boson may be able to make that change. Then we posit a new property call "emotispin," without a field, changed by photon reception. But this notion then suggests a particle that already has spin but also has emotional charge, giving an emotional magnetic moment- emotional spin. So far there has only been discovered one kind of spin, and there is nothing driving an expectation of a second type of spin. So, by simplicity, we come back to just emotional charge (joy charge).

Since typical interactions are between matter particles and matter particles take up space, and are separated by certain distances in certain directions, typical physical activities take up space, that is, occur in the world of extended objects. However, the creation of rest mass, for example relies on interaction with the Higgs field, which is scalar and omnipresent and not sourced from particles shipping charges across extents of space. There is no significant distance involved, and so the Higgs interaction takes place in nonextended fashion. So too might the joy (fundamental consciousness) interaction.

It's possible that a flip in one direction is painful while a flip to the other is pleasurable. But probably not. Probably, if feeling is based on a two-state system like spin, there are two different qualities, each with a spin, so each can be reset to triggerable state without causing feeling (though the flip-back could be randomly-timed, reducing potentiation.) Sustaining pain may be possible by units flipping back and then being flipped again.

An experimental test would be to subject the brain to different frequencies of RF to evoke pain or pleasure. Are there frequencies that can directly invoke the emotional pain response (dysphoria)?

The flipping of neural membranes could be the flip, we might propose. They don't physically flip, but they do flip voltage. However, this does not look like a fundamental action; it is a complex action and not likely the direct cause of fundamental consciousness events.

We think what we know now in physics is the bottom, but it is possible that some of the "extra" seven dimensions and actions contained therein are more fundamental, with the happenings in our familiar four, a consequence.

The emotional fabric of our minds is organized of the nonmaterial stuff of the universe, mirroring the structure of our brains, organized out of the material.

The most likely possibility is that the quantum joy field is poked at all the time and does feel little bits all over (about which we are unaware), but generally not in the organized fashion in which the brain acts upon it.

Particles form or decay at certain specific energies (which characterizes them as resonances), throwing off or absorbing photons to balance the energy difference. Perhaps this, or an atom going to a lower state, can happen by emitting a senton instead of a photon.

What is different about the brain compared to all other objects? Lots of electrical things are going on in close molecular quarters of special shapes. There could be larger than usual complexes that are in single quantum states, which could emit sentons when changing quantum energy levels.

Conservation of energy applies to the material world. While it is not proven that this law applies to the joy field, that, by consistency, is the favored presumption.

With the splitting-off of gravity long ago we received spacetime- the world of extended objects. We don't know what may inhabit the "extra" 7 dimensions called for by M-theory, or what interactions might exist at that ("nonextended!") scale.

Gravitation is unusual because there is no distinct gravitational charge. Gravitational charge is simply energy of any kind.

There is a deep bridging between matter and the electromagnetic force. Matter always has rest mass. The electroweak interaction demands a Higgs interaction that gives particles rest mass, and includes the only bosons (along with the Higgs) that themselves have rest mass.

The bosons with mass have extremely short range. A nontrivial locality is needed for joy point adding and contiguity in the field. Thus, we predict a massless senton.

Here we compile the properties of the bosons:

	Spin	E Charge	Weak Isospin	Weak Hyp	Color	Joy	Mass	Stable?
Higgs	0	0	-1/2	+1	N	0?	Y	N
Photon	1	0	0	0	N	?	0	Y
W	1	+/-1	+/-1	0	N	?	Y	N
Z	1	0	0	0	N	0?	Y	N
Gluon	1	0	0	0	Y	0?	0	Y
Senton	1	0	0?	0?	N	?	0	Y
(Graviton)	2	0	0	0	0	0	0	Y

So far:

All particles with no rest mass have no electric charge, so senton has zero electric charge.

Most particles have no color charge. None of the long-distance bosons have color charge. The senton has no color charge.

All zero rest mass particles are stable. The senton is stable.

No long-distance particle acts on itself; we speculate neither does senton.

The senton's speculated qualities then make it indistinguishable from a photon.

This means either

    1. Sentons work in a different space (dimensions).

2. The sentons do carry joy charge and react with each other, whereas the photon does not carry joy charge. This is interesting because it would further promote a fusion of feeling.

3. The sentons do not carry joy charge and react with each other, whereas the photon does carry joy charge.

4. The photon *is* the force carrier for fundamental consciousness ("let there be light!").

5. The senton does carry electric charge. This seems unlikely since we've never detected it, though it would explain interaction between electromagnetic and conscious phenomena. A possibility is that they usually travel in +/- pairs.

6. The senton does carry weak charges.

7. The senton does have a little rest mass.

We cite 1, 2, 3, and 4 as the most likely possibilities: the photon and senton differ in that they operate across different dimensions, or in that one carries the joy charge, or they are in fact the same particle.

The fermions all have weak hypercharge (and weak isospin). All (or some) fermions might also have joy charge. This is one-way interaction with the electroweak force could occur.

Since the brain uses ions instead of electrons, it is possible the nucleons contain the joy charge, which means the quarks (or the gluons).

Quarks are the only particles known to experience all four forces, and the only ones to have non-integral electric charge (and the only ones to have asymmetric opposite charges).

Perhaps non-integral or nonequal electric charge magnitude has some connection with joy field interaction.

Something must carry the joy charge. There is no requirement for there to be a new particle to bear the charge, it may be carried by one of the known particles as the charge of an undiscovered force.

Properties of the fermions:

Particle	Spin	E Charge	Weak Isospin	Weak Hyp	Color	Joy	Mass	Stable?
up/charm/top quark	½	+2/3	Y	Y	Y	?	Y	Lightest
down/strange/bottom quark	½	-1/3	Y	Y	Y	?	Y	Lightest
electron/muon/tau	½	-1	-1/2, 0	-1, -2	N	?	Y	Lightest
3 neutrinos	½	0	+1/2, 0	-1, 0	N	0 ?	Y	Lightest

In the case of ions, electric charge movement due to the electron quantity, could be associated with the movement of nucleons, so the joy charge could be carried either by the electrons or by the nucleons.

So far in known physics, all the kinds of charges are carried by matter particles (fermions), and all the kinds of charges are also carried by bosons.

Though not yet seen, it is possible in principle to have a force that only acts on other bosons. It is also possible to have a boson that, once created from free energy, then only acts on itself!

The joy charge carrier(s) is probably not the neutrinos, because there is nothing neutrino-ish about brain activity. All 3 other fermions are possible, all of which carry electric charge, and weak hypercharge.

The candidates to carry joy charge at this point in the argument are Higgs, W, Z, photon, quark, and electron. The simplest case of how electromagnetic behavior (as in the brain) could cause joy behavior, is where the same particle carried both electric charge and joy charge. That would narrow the field to W, quark, and electron. The W is involved in neutrino generation and absorption and nuclear transmutation, which appear to have little to do with brain activity. This leaves the charged matter particles: electrons and quarks. We provisionally conclude that it is likely that one or both particles carries joy charge. A little more complicated solution, is that whatever boson mediates the electromagnetic force (the photon), carries the joy charge (which was one of the scenarios we marked as likely earlier). As a plan B we add the photon back to the list: electron, quark, or maybe photon. Intriguing about this, is that photons take up no space and do not make extended objects. Everything on this three-item list is either electrically charged, or the force carrier for things that are.

Now, if sentons were to also carry the joy charge, they would interact with each other. Bits of fundamental consciousness affecting each other is necessary for a successful physical theory of consciousness, given our personal knowledge of pain and pleasure adding into one experience. We have postulated this happens because of the joy force field. Sentons carrying joy charge would be another route for bits of consciousness to fuse into an emergent whole (and perhaps enhance time-persistence). It is possible that both things happen.

8.7. Further Conclusions

The particles that carry joy charge- the "targets" of the joy force interaction- are probably the electron, quark, and/or photon.

Sentons might also themselves carry joy charge.

Predicted properties of the senton:

Stable.

Spin: 1. We need extended locality for value addition in the field. Therefor is not spin 0 (scalar boson). All other bosons have spin 1 except the postulated graviton.

Weak hypercharge: 0

Weak isospin: Y

Mass: No

Color: No

Electric charge: No

Joy charge: No?

8.8. Perspective

All we have seen here carries the implication that fundamental consciousness is part of the Universe itself and does not require human existence, in the same way as motes of dust exist. Organized compilation of these basics to make things like DNA and what humans would call consciousness do require organisms like us, however.

All other known physical phenomena are describable as particles and radiation, interacting. For example, electricity is the flow of fundamental particles with electric charge. Consciousness as a physical phenomenon is in some way related to particles.

Bio-evolution has exploited all macroscopic forces, and it has exploited joy, a.k.a. fundamental consciousness.

In the process of evolution, the forces of gravity and especially electromagnetism (which explains both electricity and most of chemistry) have been discovered and exploited heavily. As brains and their minds have become capable of ever-wider scope, and have become social and cooperating, and have become learning- and knowledge-oriented, emotional feelings have been discovered, to assign value and to paint a useful impression as to how things are going overall, to supplement the very small time and space scales in which the rest of the brain operates.

Consciousness is useful to promote successful gene replication in a competitive environment. Because consciousness makes for both smarter individual brains and teamwork between them, consciousness aids in the successful replication of the genome, supporting it in certain niches of the ecosystem. Consciousness will be retained, and its use, improved.

It is the field that feels, not the brain that tweaks it! If it is pulled in lots of small and different directions at once it feels no net emotion, and a patch of consciousness is created in it that is rough-surfaced but not much distended.

If artificial means perturbed the field like you do, your such clone would continue to feel- to *really* feel.

Just making each bit feel at the same time, independently, without being members of a field, still would not feel unified. The field itself feels a big net pull. (It just so happens (?) that this is similar to what happens on a neuron's membrane- little impinging bits affecting the summation).

This also means that technically our consciousness is an "extended" thing (somewhere)- it is just not made of matter. Different parts of a physical brain (or many brains) can poke at different parts of the joy field.

This consciousness field, from a modern physics perspective, is like an ambient atmosphere that the brain works on. As far as the disinterested brain is concerned it is an outside resource that it is just exploiting, because it helps deliver the right practical results.

If you move a cell phone, it perturbs a different place in the electromagnetic quantum field but does not notice this. In the same way the brain uses the ambient electric field as the head is moved around.

The key is a particle, or cascade of particles, that together contain both electric charge and joy charge. The movement of the electrical charge would then move the joy charge as well. Why is the brain doing things with this field that other things aren't? At the lowest level, it isn't! The difference is that the brain is organized.

A neural signal transmission has an electrical pulse traveling one way surrounded radially by electrical flows. Perhaps electrical charge itself is always messing with the joy field but the brain is more consistent and structured in these interactions. In addition to neural signals, the brain also has tightly controlled synapse gaps, and neurotransmitter vesicles opening and closing.

All phenomena are represented by- are made of- particles and their interactions and movements. The carriers of joy charge or the force carriers (bosons) for the joy force could either or both be particles we already know.

Conscious energy exists, therefore conscious power and entropy and possibly radiation. Conscious energy does work. The bosons likely have momentum and position.

It is fascinating that an unknown mechanism causes collapse of the probability wave function to a specific event. It is possible this action correlates to a speck of fundamental consciousness. That does not necessitate that consciousness causes the collapse; it may be the other way around.

The simplest (or nearly-simplest) that can work is probably the correct answer (inductively-derived axiom).

It is simplest that all phenomena that occur in Nature are of Nature.

Ergo, probably all phenomena that occur in Nature are of Nature.

Consciousness occurs in Nature.

Ergo, probably consciousness is of Nature.

All phenomena of Nature are describable in particles (inductively-derived axiom).

Ergo, probably consciousness is describable in particles.

8.9.   Review

An atom has no net charge and common objects have no net charge, because the charges within cancel out. Positive and negative still exist and bubble and froth.

The feeling of each joy quantum is small. The field adds, with each charge. This is where the unity of emotion comes from. This, and the co-stimulating-of the-field loops in the brain. The connections make them all fire at the same time; the adding together in the field is what produces the strength. I am the composite perturbation of the M-theory field. As my head moves through the field, the location of the perturbation of the field changes, but it is still Me, as with any other case in our physics, such as moving a current in a wire to different places in a magnetic field.

Where the perturbation is random there is not net pain or pleasure, such as in cognitive neocortex, which runs along the top of your cerebrum. Where it is of the same sign in the same direction, in the older brain regions along the bottom of your cerebrum, you feel pain or pleasure. If the simultaneity, in locality, is reduced, I do not feel much pain or pleasure.

There apparently is macro-scale locality in the consciousness force- cortex is large and spread out for the key individual cell types, whichever they are. This means the force is not very short-range like the nuclear forces. The simplest assumption is the inverse-square law of electromagnetism. One might speculate that the electromagnetic interactions in the brain occur per inverse-square as they always do, and then trigger a short-range joy force interaction at each point along the electromagnetic interaction pattern. But then the points of feeling would live separately and not add into a unitary feel.

Per Occam's razor, the simplest view is there is a single consciousness mechanism at bottom. Therefore, it must be a single positive/negative, or perhaps one positive and one negative, but that's all. Value adds with sign. There is no positive or negative *direction* if there is universal space for consciousness, so there is a positive and negative charge.

Our brain organizes the raw consciousness of the universe, just as it organizes ions and molecules. Fundamental consciousness is a brute fact of our world, like electricity. It is a force of nature. Like all such brute facts, we know consciousness is possible, because it exists, and that it happens to be actual in this anthropic universe.

8.10. Inter-force Mechanism

It is known that there is a connection/interaction between consciousness and the "physical world" because brain causes mind. We know consciousness interacts with matter (or force) particles in the brain. For example, we know that reading printed facts can evoke emotion.

For the brain to do what it does, consciousness must be connected, in some way, to the Standard forces. *Standard* nuclear activity does not happen much in the brain nor does gravity seem to matter. The connection point to the consciousness force should be to the E-M force (or to the quantum reconfigurations such as when molecules bind and split).

Feeling is a fundamental action, so in physics, an interaction. The creation, travel, and/or absorption of the feeling boson (senton) is when feeling happens. Matter particles can in principle carry sentience charge (joy charge), and the senton may carry electric charge, or an electric charge-carrying particle also may carry joy charge. Though we haven't seen it yet, there could be bosons that act only on other bosons, with the charge unrepresented in the world of matter, i.e., the world of "extended objects" that take up space.

By way of the joy boson (the senton), the joy force operates on any particle with joy charge. These are either existing particles for which we didn't know they have joy charge, or new particles. The boson may not have any joy charge itself, but it may have electric charge. The senton is the only new particle needed if existing particles have joy charge (and it is conceivable that the photon is also the senton). If electrons- or protons or neutrons since all the electricity in the brain is ionic- carry joy charge, their movements might emit joy waves as well as magnetic waves.

Moving ions seem to be at the heart of the matter. Unlike conventional electricity, ions have nuclear forces at play as well as EM, and thus the scale of seven "unused" dimensions. Ions have constituents that engage with all the forces. Electrons have no strong force. Free electrons in current may never get close enough very often to other needed particles to participate in weak interactions. Ions are far more massive. Electrons are point particles (or nearly) and ions all have size.

Joy field Interaction gives particles their emotional kick- perturbation of the joy field causes feeling. (Perhaps joy waves travel circularly, staying in the same place because its dimensions of freedom are so small.) Sentons themselves may carry joy charge and interact as a group with other local sentons.

If fundamental consciousness is directly caused by the specific organized movement of a charge we already know, that charge is probably electric charge; electric flows that are organized in some way might cause consciousness. Perhaps charge *acceleration* produces feeling. Or current *increase*. These appear in the ion transmembrane flows of neural firing.

Another possibility is the molecular throwing-off of bosons. Energy could be lost as a senton instead of a photon. Emission of a senton could feel bad, and its absorption, good (or vice versa). Any consequence of electrical activity that encouraged a quantum state change like this, could be the link between the forces (perhaps in both directions). This case is interesting in that it is a quantum wave function collapse- potentiality to reality. Consciousness is the experiencer (some would say here, the determiner) of reality.

Ultimately, we need an interaction link from moving ions to the consciousness force. It is not certain which one is correct, but it is clear that there are several possible pathways for an interforce mechanism. The net effect, in each case, is ion control of the joy field.

8.11. In and Out

In the previous section, we looked at how "normal" physical operations could cause consciousness events. To be useful (which it is), once feeling occurs, it must have some way to then affect the neural decision making. In practice, pain and pleasure are motive and determinative. This stems from one of two theoretical possibilities.

1. Pain and pleasure themselves have some kind of direct energy effect on the functioning of some neurons, or
2. Pain and pleasure are physical phenomena that serve the role of summary scores resulting from a lot of neural actions. It so happens that these are feelings, and these feelings are *cognitively* observed by circuits that mechanically turn away from the current behaviors that cause pain. That the pain phenomenon ends up thus avoided and the pleasure phenomenon is pursued, is a happy accident of this universe. In a miserable twin universe, the reverse would just as well serve the biological purpose. Just as with the illusion that our decision-making (such as to move a finger) flows from our completed conscious reasoning, it is an illusion that the subjective negativity of pain "causes" us to do something.

In practice, both mechanisms may be involved.

The simplest explanation is that electromagnetic things affect feeling (fundamental consciousness), and feeling affects electromagnetic things, rather directly. In other words, the brain not only outputs to the joy field, but inputs from it as well, directly. Then pain and pleasure do motivate and determine, directly.

Again, we remind of the Anthropic Principle. While it may seem incredible that pain and pleasure would so conveniently exist at all for such evolutionary exploitation (regardless of whether used as in possibility 1 or possibility 2), it may be that in a very large number of other universes they do not/have not/will not. Something as special as feeling is needed for things like us to emerge (and other such things could seem just as natural or inevitable to other beings). All that is needed here for intelligence is some mechanism by which to attach summary value. As a single unified cosmic force splits into specific, separate forces, the universes that physically evolve something that serves as does feeling, will bear their "feeling" fruit.

By which exact mechanism feelings affect neural operation is not clear. But the interaction in this direction, "out," probably in some way resembles one of the possibilities discussed in getting from moving ions to feelings, that is, "in."

8.12. Approaches to Experimental Confirmation

A changing magnetic field produces a current, a changing electric field produces magnetism. It is possible the electromagnetic-to-consciousness mechanism is fundamentally bidirectional. Consciousness would then directly produce electrical effects (and these could be observed). If production of feeling is reversible, feeling may be able to produce electric current. It is interesting that there is great axon parallelism in the brain in the same direction, and myelinization.

If it can be predicted what conventional qualities a joy-charge-interacting particle would possess, we can search for it. It likely is weakly interacting or it would have been discovered already (this makes candidates of dark matter and dark energy).

If neural tissue of two brains are placed very close together, are one's high or low feelings affected by the other's? Can one detect intense feeling activity in the other? Does intense feeling in one change any neural firing in the other?

A nonbiological experiment would show a contraption acting so as to stop pain or continue pleasure. If a system were built that could truly equally go either way but favored one direction of adjustment, it might be conscious.

The joy field may be impinged upon in disorganized fashion by the nonfeeling neurons of the brain. This would have appeared in the evolution toward well-tuned feeling mechanisms. This may result in a background of consciousness noise in the brain. There may be some way to experimentally detect this. There may be more of it, or it may be easier to get at, than the feelings. Note that the cosmic background radiation went unrecognized for a long time.

Perhaps Transcranial Magnetic Pulses work for reasons other than, or subtler or more elaborate than, theorized. Further experimentation with TMP may be fruitful.

Feeling is a consequence of location and/or movement of joy charge, and like the other charges, may be involved in quantum complexes. It is also conceivable that consciousness-related decay events occur.

We could look carefully for known particles being thrown off by animal brains. There has never been such a search for neutrinos in or out, for example, and there is a known connection to the electroweak force.

We might discover additional emitted photons, of certain frequencies, caused by quantum state changes like re-conformations or spin flips, in an operating, feeling brain. We could look for a preponderance of certain frequencies during painful experiences.

What would we find by putting an animal brain in a bubble chamber or other such detector? What particles might be leaving our brains without our knowledge?

There is a lot of energy changing hands in the brain. Apparent violations of conservation of energy (or perhaps momentum) could expose interactions involving the joy field.

8.13. Observations and Open Questions

Does fundamental consciousness have a close relationship to one or more of the remaining large physics mysteries. Is it related to collapse of the wave function? Is it related to gravity? Is it related to dark energy or dark matter?

A reason gravity and consciousness could be linked, is that (according to the Max Planck Institute), hidden dimensions (where consciousness may be operating) could have an influence on gravity waves.

Where does consciousness live? All forces play out in space, this one probably does, too. Where do things live in this universe that exist but are invisible? On scales smaller than the Planck length? Coiled inside the "extra" seven dimensions? Or are these "things" purely a movement or effect of something else that is already known to us?

We may take the prediction of other dimensions by M-theory as a possible clue, that there are things happening there of which we have been unaware.

The extra dimension(s) might not be small, just somehow different (as time is) and contain different things. The orthogonality of consciousness, living in those dimensions, could explain its invisibility.

There could be another sort of "right hand rule" for electromagnetism or other force, describing effects in those dimensions. Are there forces which span from our four to those seven, and are there others existing only within the seven?

A fascinating consequence of our consciousness theory is that at the theoretical Big Bang the consciousness force would be unified with the other four, and just as real and basic from the very beginning as they are. The consciousness force split out at some point after the Big Bang. A single mega feeling would not require space. but would require time. It is interesting partly because it does not need to be relative to any other thing to be.

Cosmology also means that matter and consciousness are fundamentally the same stuff. Ultimately, in a pattern that is recurrent in Nature, it is neither that matter creates fundamental consciousness or that fundamental consciousness creates matter, but that they are two sides of the same coin, emerging long ago from the same stuff.

Feeling is an essential property of the universe. In fact, it *must* be there to exploit in this anthropic universe. Probably, it has been there as-is for the almost whole life of the Universe.

Probably, fundamental consciousness happens a lot everywhere (and is not particularly special in ambient form). What is special is how brains organize and concentrate it into what we then dub as consciousness.

It is also conceivable that fusion of feeling, or the interforce connection, is actually because of a joining into quantum systems, like atoms in a molecule. This would not appear at first to be the case because of the physical scales in the brain, but we are not certain where the joy force resides and acts.

Atoms and molecules (and therefore ions) can absorb and emit photons and change state. That action is not directly related to the molecule's electric charge profile, simply that they are able to receive the boson. Similarly, a change in shape might sometimes emit a senton of emotional energy instead of a photon of electromagnetic energy. A question in that case would be what would make one of those emissions happen rather than the other. If that were understood it might be possible to make a Joy Emitting Diode.

It appears nonconscious cortical neurons might also perturb the joy field but not in the same way- random noise. That consciousness activity might serve to keep track of where attention is and/or aid short term memory, which are non-emotional acts. That perturbation of the joy field would create just activity rather than organized, directional activity. In contrast, the limbic perturbation is perhaps shaped to push in the same direction.

If a system were cooled enough, there would be little or no pain, not just because biological processes would stop, but because the system would not contain the necessary energy. Feeling has energy.

The distance scale involved gives us a possible clue about the size of the joy force constant and/or the nature of the strength equation. (The scales suggest the possibility of an inverse square law. From this can be speculated a force constant value range.)

Comets or asteroids could have provided us particles that have consciousness charge. They could have come from anywhere.

9. Where is the Site of Action in the Brain?

Drugs, anesthetics, electrical stimulation, surgery, and injuries all show that the brain causes consciousness. As to consciousness, where is the site of action in the brain? This question is actually two. The first is: Exactly where in neurons is consciousness being "made?" The gross version of the question is: In what parts of brains are these neurons located? We are most interested in the first question. In this chapter we also touch on the second.

We can contain what we don't completely understand, characterize and compartmentalize it- in this case, what we are calling fundamental consciousness. Once that is in a box- a component- the rest of the elaborations needed to make what people call consciousness can be understood on their own, using it.

Engineering and engineering understanding is replete with this kind of thinking. To deeply understand transistors, one must understand them per quantum physics. Most electronics engineers are not quantum physicists, but they produce extremely sophisticated circuits using the described transistor phenomenon.

As to who or what it is that hurts, the answer is "a nerve cell," in the sense that it is the action of a nerve cell that causes each bit of pain. At the lowest level the pain is freestanding (it does not need to be experienced by something else), but it is *caused* by the cell. Individual cells thus hurt. Pain inhabits the space where the brain is because pain inhabits the space where these neural cells are.

Some brain reaction to a stimulus must cause internally a response which is then freestanding, and no longer about the stimulus, at the lowest and smallest level- a freestanding feeling. The fact that pain and pleasure at a human consciousness level are asymmetrical to us- that they seem to be not just opposite but separate- does not mean that they so are in in Nature.

The most defining characteristic of neurons versus other cells is the generation and communication of neural spikes. And the most conspicuous action in this firing is the travel of charged ions. Furthermore, the charge travel pattern is highly complex, unusual, and dimensionally precise among neurons of the same type. There are charge flows at right angles to the spike direction, and these are in a circle, rushing toward each other as though to collide.

It appears that some neuron firings cause pain while others do not. Presumably only certain kinds of neurons cause pain. (Physical pain is especially good at causing what we mean by pain here (dysphoria) and is easier to track than mentally-originating pains.) The gross structure of the brain also suggests the speculation that the horizontal, radial, nonparallel connections in the cortex don't matter for base feeling and (some of) the ascending/descending cortical tracts do.

Pain discourages certain action choices, but *how* are pathways that cause more pain discouraged? If there were direct cognitive feedback at the neural level that paralleled the experience of pain, there would be no need for the feeling of pain, as it would not be doing anything. Suffering that serves no purpose would violate Occam's Razor, especially in complex, evolved organisms. At our level, pain experiences can be elaborate and nuanced. It is unlikely to all be an epiphenomenon.

While pain technically could be an epi- or vestigial phenomenon, it has been evolutionarily retained and elaborated upon, so apparently it has utility, which means the pain field (the joy field) must feed back somehow to the operation of neurons. It is important to understand that while cold, cognitive things can *effect* pain or pleasure, which can then *affect* cold things, the *"feeling"* is *happening* rather than being *experienced* by anything else.

Pain and pleasure are sense-able by the non-emotional circuits, so that the latter can be steered by the former (in its totality, this is "will" in action). Whatever physical mechanism in the brain does that sensing could be used to sense the pain field experimentally, or for technology.

We see that it may well be that something about the ambient joy field affects the local physics applicable to neural spike generation and propagation. For example, it could be that an increased ambient pain field slows speed or acceleration of moving charges. It may even be that Maxwell's laws or the electroweak theory can be generalized to include the joy field.

A spike is a potential difference change at the neuron membrane, but rapidly (and more) instead of the gradual change of neural stimulation. If a spike, which is an intense and rapid change of charge gradient, emits sentons, perhaps the membrane somewhere is well-suited to receive sentons. As with photochemistry, particular molecules are likely receivers.

An onslaught of sentons could erode a neuron's inclination to fire.

9.1. Unified Feeling from a Neural Perspective: Two Theories

The unitary-ness of consciousness, vs. independent spots of feeling on their own, is one of the chief (most difficult) phenomena to explain about consciousness, especially if we accept (which we do) the notion that the unitary feeling is made, at bottom, of these independently feeling spots. Why do they not just feel individually as solitary scattered motes (as may happen across the Universe)?

There are two basic theories we now examine of how these independent spasms fuse into a whole- two ways in which the individual spots of feeling may join. One is by the cyclic regeneration of each individual spot of feeling by the linked neurons that are creating them. The second is by their locality and attendant co-influence in space by virtue of the behavior of force fields.

9.2. The Field Theory of Fusion

Let us turn first to the theory based on the nature of physical fields. A charge produces a field, a second charge produces the same kind of field, together they produce one net field (with a more complicated intensity distribution). Each charge has an effect (force) upon the other, which each one experiences, and by nature of each particle being there, there is energy in the field itself, and there is change in this energy field as the particles move, and any or all these things may correspond directly to feeling.

The pain experience that we have is because so many pain spots turn on at the same time, and they effectively fuse (not just into one perfect particle of pain but a complex with structure or shape owing to its graininess and nature as an assemblage).

9.3. The Network Theory of Fusion

In the second view, a neuron that is causing a spot of feeling is also ultimately connected to and stimulating another neuron which is causing another spot of feeling, and that neuron ultimately is also stimulating the first. Or, in a variant, the same observed physical phenomenon independently but consistently stimulates both of these feeling-creating neurons, and that external reality instead is the bridge.

When "I" feel bad these fundamentally independent blips of pain occur all at the same time, and each one pain is effectively partially causing the others. These granules of pain are a simultaneous mob, and the cognitive (dispassionate) observance of pain outputs to more pain neurons. These cognitive links provide the fused mob.

9.4. Fusion of Human-Level Feeling

Pain, *at the human (or vertebrate) level*, is something that I am *aware of* when it happens. Awareness requires an observer.

Fusion into a whole of recognition is caused by a network of cognitive circuits- the observer. The complete pain experience at the *human* level includes a knot of being cognitively aware of many granules of hurt.

In the "hearing" of all that pain together we are aware of the pain- but objectively. A large cloud is happening in one place. But *we not only observe it as one, we feel it as one*.

We *are* the pain and pleasure (as well our cold observations). In the same general place our cognitive circuitry observes what is going on, there are large clouds of feeling in our heads.

9.5. Rejection of the Network Theory of Fusion

The spots of pain or pleasure fuse together because they already are fused together- they are each a ripple in the same pain field and they each add to the field. (As with all other known fields, it neither spreads all the energy infinitely with equal intensity nor contains it all in one spot- there is quantitative locality.)

We know this because simply wiring connections between spots of feeling cannot make them *feel* as one feeling- they would be independent feelings still. Feeling is a base, fundamental event, without any separate experiencer. Introspection tells us that these thousands of base feelings that must be happening, are feeling not just simultaneously or in synchrony, but *together*. Otherwise, a hundred small pains would not feel like one large one, they would just be all small ones. To be in a lot of pain, the pain must *add* somewhere. That fundamental somewhere is the field.

Every large dynamic quantity occurring in nature, every such intensity, turns out to be the sum of fundamental small values. While we don't need it, this is yet one more piece of evidence that feeling is a force of nature like the other forces, represented by small quanta that add to produce large intensities.

The network theory we have described is likely also at play in human minds, but it is neither sufficient nor necessary to create the basic fused feeling. In contrast, the field theory of fusion is both.

The neural network "feels" the unitary feel, not directly itself because of something magical about the networking, but because of the direct ("out of band") interactions between the independently feeling spots. Technically the network itself does not feel. What the network does is make the feeling spots happen, close in space and time.

If another head could overlap with yours without damage, which would make the joy events all very local to each other, then you two would become one, feeling-wise. And yet your cognitive would remain separate.

9.6. Neural Networks

We have seen that a network is not enough- a force must be introduced. We have also seen that a network is not needed to explain the basic mechanism of fusion of feeling, and it cannot do so. This all leads us to the question of whether a network is necessary to complete the actual basic feel of our actual brains. The answer is Yes. Some kind of network must exist so that, at the least, many spots can be turned on at once, from many neurons of the right kind spiking at the same time.

In principle, the nature of the links (neurons) interconnecting feeling neurons may be purely cognitive- that is, no bridge neurons existing of the feeling kind. In fact, in principle, there may be no direct feeling-neuron-to-feeling neuron connections of *any* kind. These interconnections may exist, but what *must* exist- at a minimum technically not a neural network at all- are connections to multiple feeling neurons, stimulated by the same happening in the external or internal world. This is not for the purpose of fusion of feeling- directly- but to provide a multiplicity of spots of feeling that will then fuse in the field.

Note that single neurons synapse on many destinations. This alone could cause the simultaneous stimulation of multiple feeling neurons (and multiple loci of interaction with the joy field for each such neuron) and is probably partially responsible. This kind of distributing circuitry could have been originally developed by nature for cold analysis or activation purposes, and conveniently exploited by evolution to create fused feeling. Most likely, the cognitive and feeling portions of higher, fused emotion evolved together in these networks.

Striking, is how separate the emotional and the cognitive centers are in the brain, even at a very gross level. This suggests their fundamental differentness. The most modern areas of the cortex are guided by feeling to make sophisticated decisions supported by extensive analysis. This is done at both the microscales and macroscales of the brain. At the micro level, feeling guides neural conclusions about small details. Neurons further up the processing chains will be guided in making higher-level decisions, with the same feeling mechanism originally used. It is possible that with evolution, smaller scale deciding, which had been all there was, has been turned over, perhaps necessarily as the brain got more complex, to cold cognitive evaluator circuits. Top-level choice motivations, in any case, remain very emotional.

9.7. Feeling Guiding Thought

Now we have the discussed the causing of feeling. What about the direction back, to our feeling affecting decisions? How is the guiding of cognitive processing done? (And once are evolved circuits that work, could feeling go away?) Having touched on the question in our exposition on the joy field, we revisit the matter here.

Neurons might also *sense* the joy field in some way and so be affected by it. Or instead, quantitative value indications- cold numbers- from the same clot of neurons evoking feeling spots, could be provided to the decision centers, which would switch on their basis in a calculating fashion. Why, if the latter is the case, do we avoid "averse" sensations? It seems quite a happy coincidence if we just happened to be *wired* that way. Why then should a hot stove be painful, if we would make the decision just from the wiring anyway? If that were the case, the pain we feel would be simply a physical side effect without functional purpose for the calculation proceeding at that level. But even in that case, the generated pain or pleasure might be accumulating and guiding higher-up processes.

It has been noted that more automatic operations of earlier organisms have become corticalized in humans. To be specific, they have been highly *neo*-corticalized and woven into the analytical and action execution portions of the brain. Those portions would come under the influence of the feeling portions of the brain.

It is unlikely that feeling is only an epiphenomenon of a functioning brain, a thing that just happens to exist in this universe, and also happens to happen, so pungently, in operating brains. Feeling has utility.

For feeling to have any utility, it must be sense-able in some way by neurons, such as by altering a part of the complete physical context in which those neurons operate.

The joy field is used for net value calculations (warmer vs. colder), on a relatively large spatial scale. Its value is in providing value, as an additional field exploited solely for its own behavioral features and separateness from the related electromagnetic processes. In that sense and to that degree, feeling is an epiphenomenon. Feeling is a physical process that serves an evolutionary purpose in this anthropic universe. While, we have always found ourselves and our consciousness to be terribly important, feeling is just another process in the universe. (But it is ours, and there is nothing that is more important.)

The adding and fusing occurs in the field, so the field- feeling- is (or was) necessary, and the stimulation of the field is (or was) necessary. Either the adding is nowadays also happening in circuitry and the field perhaps no longer needed, or the field must itself be sensed to gain access to that adding, for the distributed adding does have necessary computational value in decisions.

It is theoretically possible that the summing in the field was never (or not lately) used by bio evolution, that the summation of value is done coldly by neural nets, and the gross summed feeling is an accidental consequence of all those neurons closely packed. Neurons are fundamentally summers. This would also remove the need for any field. But, again, it is unlikely that feeling, which exists in this universe, serves no purpose in Mind.

An (the?) advantage of adding in the field is that it gives a macroscopic score, without needing a network to evolve to completely do the same thing successfully. It is a simpler mechanism, already extant and exploitable.

It could be that the joy field provided the mechanism to give a kick start, and neural networks have themselves taken over more and more from there. In fact, that is very plausible for the lower and earlier levels of processing which are now far below our human consciousness. But the takeover is not complete. The exquisitely able human animal features both sober, deep reflection, and passionate preference. Thus, none of this discussion removes the necessary existence of a joy field.

The palette of contributions by the joy field that it makes available to evolution are

- Separateness and orthogonality from electromagnetism
- Its own laws of action, such as with signed-ness
- Adding and fusing
- Taking up none of "our" space
- Possibly brief persistence (memory)

Again, (even if only at an individual spot level) for the joy field to be useful, it must either be sense-able, or it must affect the overall physics context in which the brain is operating. But the latter case amounts to the same thing; it still means the brain can sense at least the broad state of the joy field. Whatever interaction exists from electromagnetic activity to joy field activity, it is accompanied in the complete laws of physics by some interaction path back the other way.

It may be that while the poking at the joy field is in pinpricks, the sense of it is blunt. That would work just fine for its purpose. Strictly, it is not known whether the joy effect on the brain networks (that is effective) is from single points or larger-scale summations. That is worth speculating on, but it is a secondary question of detail. If the joy field is used for point by point utility, it could provide persistence (memory) for microscale brain operations. Note that electromagnetic effects are used in the molecular brain both very locally in the brain without using field summation, and additively in terms of net charge and thus voltage difference across the membrane.

It is unlikely that feeling is totally just a side effect, as it would then exist in our universe as another specific force for no anthropic reason; possible, but less likely. The most natural to expect, is that if electromagnetic action can cause joy field effects, then the reverse is rather directly also true. These reverse actions in some way affect the brain's actions- they influence neural network activity and they will correspond to a fundamental physical interaction. Simply issuing sentonic energy would reduce the energy of the donor system and thereby change its activity. Simply receiving them at another site would do the reverse. The technique could give the brain another means to cable itself. (New fields and quanta of a non-conscious nature could do the same.)

9.8. Macro-Scale Facts and Conjectures

The brain is made almost entirely of: capillaries, fat, glia cells, extra-cellular liquids, mobile blood cells, and neurons. Animal-level consciousness has been well-correlated by experiment with widespread neural activity- in other words, neurons firing and stimulating each other- and does not seem due to one small center of the brain. While it is possible in principle that the neural activity does something to the capillaries or blood cells or fluids that causes consciousness, there is no compelling reason to prefer to look that far afield from the neurons. (It is possible that the glial cells or the fat are the site of action for consciousness. But even if they are, the odds are that the "mistake" of assuming the neurons are the site of consciousness will not likely lead us long and far off the trail.)

Interestingly, vertebrates are the conscious ones (we think) and only vertebrates have myelinated axons.

The nature of the neural populations of the known pain and pleasure areas should be studied for differences from other populations. Knowing which neurons directly cause pain or pleasure would be helpful. Noticing which are similar between both the pain and pleasure centers might also be helpful. And there might be other clues hiding in the contrast between human pain and pleasure mechanisms. Pain has receptor cells for it. Pleasure is always an interpretation; there are no pleasure receptor cells.

Fewer neurons firing seems to have weaker effect. Everything indicates each one of the right kind of neurons results in a single bit of emotion.

The signed emotional charge view suggests something is separating negative from positive feeling-charged particles. This could be done by neurons. If one's head were sprayed with only positive joy particles, one would feel happy.

Different neural tissues, some of which appear to be feeling and some of which do not, contain different types of neurons. These neural types include:

- Pyramidal- primary excitatory neurons of cortex. Glutamate excites, GABA inhibits.
- Stellate- receive excitation from thalamus, send excitation to layer II/III pyramidals in same region. They appear as excitatory spiny or inhibitory aspiny types.
- Spindle

Pyramidals are the "work horse" cells of the brain, and allow communication between brain areas. Pyramidals come in 3 types: RSad, RSna, and IB. In response to 400-100 ms current pulses, RSads produce one output spike at a time, RSnas produce a train of output spikes, and IBs produce 2-5 rapid output spikes.

Stellate neuron processing is in all different directions, while pyramidal outputs tend to run in the same direction.

Emotional consciousness is "smell consciousness," which developed heavily with mammals. Cognitive consciousness is visual consciousness, which exploded with primates.

Human consciousness requires only

1 cerebral cortex (for awareness)

2 RAS of the brainstem (for arousal)

It appears that consciousness occurs in the cortex. Therefore, we should study individual neurons there, and in particular, those from emotional areas.

9.9. Cellular-Level Facts and Conjectures

Macroscopically and microscopically, some areas of the brain show no sign of consciousness. This suggests certain neurons and not others have evolved conscious ("feeling") capability. To our thinking, that means they have evolved (nontrivial, nonchaotic) access to the joy field.

We note that if we look at limbic brain tissue as pain/pleasure-related, we find that it omits the stellate cells of cognitive tissue. This suggests that the feeling neurons are pyramidal.

The newest cortex is not only guided by signals from emotional centers, but a primary acceptor of dopamine neurotransmitter itself. But we note that it is also possible that the "feeling" cortex just provides stimulation into closely packed subcortical nuclei that do the actual feeling when stimulated.

We could look to where opioid receptors are distributed. The answer is, "widely."

The neurotransmitter acetylcholine has little to do with emotion; those neurons we can pass over in our search.

During REM sleep, lots of acetylcholine is released and *no* histamine, serotonin or norepinephrine. GABA (inhibitory, except excitatory from chandelier cells) and orexin promote wakefulness. Since sleep can be very emotional, it provides another avenue for research into mechanisms of feeling, such as by noting what is the same and what is different from waking, emotional experiences.

GABA works at medium spiny cells.

Membrane flipping seems most likely to correlate to fundamental consciousness occurrence, but then why are only some neurotransmitter systems involved in consciousness? That is probably based on the type of neuron that sends/receives a particular neurotransmitter, for example pyramidal versus "interneurons."

High-level function of neurotransmitters can be inferred by the actions of drugs:

Cocaine and amphetamines inhibit dopamine reuptake.

Ethanol, niacin, barbiturates, and valium stimulate GABA receptors.

"Smart drug" racetams modulate glutamate and acetylcholine.

Aniracetam and piracetam enhance AMPA reception, like glutamate.

Ethanol and PCP reduce NMDA reception.

Surveying the whole structure of the generalized neuron contributes clues for the site and mechanism of action. Also, we know some parts of the brain are very emotional and others are not; we can also look to what is different about those brain parts and those neurons to find the mechanism of base emotional feeling (fundamental consciousness). Neuron commonalities and differences are:

- Membrane flip
- Neurons drop potential difference abruptly, in a traveling ring down a tube. This is a very unusual and distinctive thing.
- Axon length and type! Sustained flipping sequence provides sustained feeling?
- Myelinization
- The neural gap (synapse) is maintained and precise.
- A process happens at the end of the axon to cause *possible* release, probabilistically.
- Vesicle opening at the membrane
- Dendrite length and type
- Receptors
- There is also a cytoskeleton running through the center, made of microtubules.
- Organelles we haven't previously noticed?

We assume no prokaryote is conscious. Eukaryotes have many organelles. One of these may be involved in consciousness.

Before moving on let us note that there is one other type of cell in the brain suited to a general behavior rather than one from a precisely structured network- the glia. It is conceivable they are involved in mind and that has been speculated. A new speculation here, is that they could sense, bluntly, the joy field, and, being wrapped around axons, have an effect upon neuron activity.

The cellular-level mechanism associated with fundamental consciousness should be something that is characteristic of neurons, different from other cells and very precise. A candidate is the unusual 3-D ion flow pattern. Another would be something about the process of the initiation of a spike from the axon hillock. Another is the rapid potential change between the terminal bouton and the precisely distant membrane of the next neuron. Another is any of the unique molecular re-conformations that might throw off or absorb a senton rather than a photon, or a photon of a specific, activating frequency.

Ion channels (ion pumps) exist in all cells, even extremely primitive ones, but *voltage-gated* ion channels exist mainly in neurons. Further, while some cell ion channels transport naked protons, larger ions are used for neurons.

Special to brain cells (and receiving muscle cells) are voltage-gated potassium channels, voltage-gated calcium channels, and voltage-gated sodium channels.

Whereas neurotransmitters are complex molecules and there are many of them, there are only three ions usually used for the transmembrane ion flows. Two different positive ions are commonly used (sodium, potassium), mostly only one negative ion (calcium). Calcium is closely related to neurotransmitter actions, sodium to spike creation and propagation, and potassium to repolarizing after spikes. When a spike occurs, sodium ions flow in, are then transported out, and potassium ions then flow out.

Anesthetics (drugs that remove consciousness) of very different molecular form are all lipid-soluble, and their effectiveness varies per their solubility, suggesting that they work via their absorption into the membranes of the neurons, which are lipid bilayers. This also suggests, consistent with our already-presented thinking, that this absorption results in a disruption of an in-some-way very precise process that needs to occur to stimulate conscious activity. This could be simply preventing spiking, or it could be subtler than that. (It also suggests the possibility of the importance of the myelinating glial cells).

Anesthetics include:

Sedatives

  Barbiturates

    Agonists of GABA receptors

  Benzodiazepines

    Enhance the effect of GABA

Many others

Reducing the brain's neural activity reduces its consciousness. The reason so many differing substances cause anesthesia appears to be that all one must do is disrupt the proper behavior that results in consciousness. Consciousness stops when we shut the neurons down.

The diversity of anesthetics suggests a disruption of a perfectly aligned cellular system for consciousness. In principle these chemicals could also absorb consciousness themselves. By joining with the membrane, they would provide an absorber. Or a barrier. (An even wider array of chemicals also removes consciousness but have additional effects that injure.)

General anesthesia disrupts transmission of nerve impulses, neural firing goes down, human consciousness goes away. Ergo, firing causes human consciousness. *Cycling* in neural networks may cause a compounding regenerative effect that is disrupted, but it is probably not the base effect.

Another striking aspect of spikes is that they are *sustained* for a very long time of travel (milliseconds), which could support resonance/regeneration in the joy field, or emit a fusillade of sentons.

Probably, neuron firing or spike transmission directly causes the conscious event. This could possibly occur at the end of the axon where neurotransmitter release is a probabilistic event, not a certitude.

It is likely that the base effect is on a small size scale. The only thing that happens in the (very constant size!) synaptic gap itself is drift of molecules and a capacitive leap of force; this does not seem like the most likely site of action. The precise gap may be for controlled neurotransmitter life cycle.

Anesthetic experiments and pathology and injury studies indicate that it's not receptors and it is firing sequences, and also that some neurons make consciousness, and some do not. The second most-likely site of action, after spikes, would be in the immediate chemical cascade in the receiving neuron upon attachment of a neurotransmitter.

While dendrites often taper, axons usually maintain a constant radius. The diameter varies by neuron type, but typically is one micron. While dendrites are restricted to a small region around the cell body, axons can be much longer. No neuron ever has more than one axon. These facts draw more attention to the neural spike.

Myelinated axons are a defining characteristic of vertebrates. Fish can learn. More white matter correlates to more intelligence.

The axon initial segment consists of a specialized complex of proteins. It is 25 um in length. The traveling ring of charge at the membrane in a spike is also surrounded by ion charge on one side and opposite charge on the other: ion flows are rushing in a traveling ring also. The physics of axonal transmission could be approximated by constructed apparatus for experimental purposes. Effects discovered could be looked for in real brain tissue.

Microtubules have been speculated to be the site of consciousness. Microtubules are very small in radius and have absolutely consistent structure and radius. They extend down the length of the axon and would experience the traveling electromagnetic spike around it, possibly inducing something in or on the microtubule. The precisely-dimensioned microtubule might be a resonator.

Overall, the evidence gives us just one good group of gross or rapid and stereotyped actions that a neuron performs, which is what we would expect to generate the special action that results in a blip of pain, and that is the action potential (spike). While the dumping of neurotransmitters consequent to a spike opens tiny circular cavities in the neural membrane, the membrane contains many channels already. The neurotransmitter molecules drifting across a cleft do not seem to be enough, binding of them to their receptors do not, and the relatively gradual stimulation building from many different inputs does not. The spasm, chain reaction, and travel of a firing do. Exactly what and where within that physical process causes base consciousness, we cannot yet say; there are multiple possibilities. But all we really have to know to start is that something about that electrochemical spike interacts with the joy field.

9.10. The Structure of Cerebral Cortex

The organization of consciousness is accomplished by the organization of the brain. Experiment and injury have shown this. This is one way we know that the complex, higher consciousness is not free-floating or somewhere else and only distantly or slightly or restrictively connected or minimally coupled to the brain. Fundamental consciousness, however, is special. The weaving of it into the structure of a mind is a direct reflection of the structure of the brain. The brain is an object that creates mind out of primitive natural phenomena.

For reference we present here an outline of the structure of the entire brain, generally from evolutionary newest to oldest:

I. Cerebrum

  A. Neocortex- 6 layers. Mammals only. (80% excitatory, 20% inhibitory neurons). Plans movement.

    1. True isocortex

      a. Frontal- action-oriented. Cognitive and motor. Contains most of the dopamine-sensitive neurons in the cortex.

      b. Temporal- higher vision and language, emotion association. Holds the semantic memories.

      c. Parietal- integrates sensory information among the various modalities

      d. Occipital- Passive cognitive. Totally about vision. >= 3 of the 50 human Brodmann areas.

e. "Limbic"- parts of the frontal, parietal, temporal lobes

2. Proisocortex- transitional from periallocortex to true isocortex.

   a. In cingulate cortex

   b. In Brodmann 24, 25, 30, 32

   c. Insular cortex- emotion "and consciousness"

   d. In the parahippocampal gyrus

B. Allocortex- 3 or 4 layers. Olfactory system and hippocampus

   1. Periallocortex- transitional zone to neocortex. Emotional?

     a. Entorhinal cortex. 3 layers. Memory and navigation.

   2. Archicortex- 4 layers.

     a. hippocampus

     b. dentate gyrus

   3. Paleocortex- 3 layers. Both granular layers II and IV are missing (so those two are likely new).

     a. Olfactory bulb

     b. olfactory tubercle of basal ganglia- reward cognition.

     c. piriform cortex

C. Thalamus

D. Epithalamus- connects limbic system to other parts of the brain.

E. Hypothalamus- Sex, fear and defeat, maternal, ...

F. Subthalamus

G. Amygdala. Only exists in complex vertebrates. Emotional reactions

H. Claustrum- "consciousness coordination"

I. Basal ganglia

    1. Caudate

    2. Putamen

    3. Nucleus accumbens- dopamine received -> motivation and desire, pleasure experience, addiction

    4. (Olfactory tubercle)

    5. Globus pallidus

    6. Ventral pallidum

    7. Substantia nigra

J. Basal Forebrain- the major cholinergic output- drives basal ganglia pleasure centers

    1. Diagonal band of Broca

    2. Substantia innominata

        a. Nucleus basalis

    3. Medial septal nucleus

K. 3 circumventricular organs- autonomic functions

II. Cerebellum

III. Brainstem

  A. Midbrain- automatic

  B. Pons- automatic

C. Medulla- automatic

D. RAS runs through all 3

The cerebral cortex is tight, dense. Most of human cortex is arranged in 6 layers, numbered beginning from the surface. About 25% of its neurons are interneurons, inhibitory, using GABA or glycine, including basket cells which are about 8% of cortical neurons. There are a few interneurons that are excitatory, using glutamate.

Layer	Contents	Input	Output	Freq
I		M-type thalamus cells; interhemispheric cortical		
II	small pyramidal, numerous stellate	interhemispheric cortical	to cortex	2 Hz
III	small-medium pyramidal, nonpyramidal	interhemispheric cortical	to cortex	2 Hz
IV	pyramidal and stellate	C-type thalamus cells; intrahemispheric cortical		
V	large pyramidal		extracortical	10-15 Hz
VI	small spindle-like pyramidal, multiform		thalamus	

Layer I contains pyramidal dendrites (feedback for associative learning and attention), and horizontal axons, and takes M-Input from thalamus and input from cortical neurons.

Layer II contains small pyramidals that project to other neocortex, and numerous stellate neurons. Cycles at 2 Hz.

Layer III contains small and medium pyramidals that project to other neocortex. Cycles at 2 Hz.

Layer IV neurons are the main reception point from outside the cortex, and they make short connections to surrounding layers. It contains stellate and pyramidal neurons of various types.

Layer V contains large pyramidals that often project out of the cortex. Cycles at 10-15 Hz.

Layer VI contains a few large pyramidals that often project out of cortex, and many small spindle-like neurons that project to the thalamus.

Connections vertically through the layers are much greater in number than horizontally-running connections.

The common cycling frequency of 2 Hz of layers II and III demonstrate a tight relationship between them. These take interhemispheric input.

The connectivity of the cortical layers is:

Layer I: input from M thalamus cells, cortex

(Newer) II: small pyramidal neurons and numerous stellate neurons, input from cortex, output to cortex

III: mostly pyramidal, input from cortex, output to cortex

(Newer) IV: stellate and pyramidal, input from C thalamus cells and intrahemispheric cortex

V: pyramidal, output to subcortical including muscles

VI: pyramidal, output to thalamus

Older (non-neo-) cortex, the Allocortex, is only 3 or 4 (or 5) layers, olfactory and hippocampus. It includes:
1. Periallocortex.
2. Archicortex. 4 (or 5 or 3) layers. Among the oldest cortex. In humans this is the hippocampal formation, which is part of the olfactory system.
3. Paleocortex. Thin 3 (or 4 or 5) layers. Most primitive, includes olfactory cortex. The two neocortex granular layers II and IV are missing. Thus, granular layers appear to be newer (and pyramidal layers are earlier).

Olfaction does not need pattern matching, because smell detects the great many chemicals that it can detect individually at the point of entry. This suggests granular layers provide pattern matching.

III and V, of the olfactory epoch, are older; consciousness is older. II and IV are newer elaborations of cortex, once conscious cortex was in place. The newer IV also takes the input from C thalamus cells, the ones that pass on vision and all the other senses aside from smell.

Cortex has evolved, to greater complexity.

- Paleocortex has 3 layers- no granular layers! (II and IV). It is the most primitive cortex, missing the stellate cells

(which are mostly inhibitory). Piriform cortex is present in amphibians, but not fish.
- Archicortex 4 (or 5) layers
- Periallocortex
- Paralimbic cortex has II,III,IV merged, thus 4 layers.
- Neocortex does not appear until mammals. If sub-mammalian brains are conscious (which they appear to be), neocortex is not required for consciousness, only older cortex. This is another major clue.

A human has 2.6 square feet of cerebral cortex (which is grey matter, as are "nuclei"- large clusters of neurons- deep within and below the crown of cerebral cortex). (More grey matter and more white matter both correlate to higher IQ.)

Human cerebral cortex has about 16 billion neurons (and thus 16 billion axon hillocks)- about 1.5% of the cells of a newborn. Typical neuron types are:

```
Pyramidal   cortex, hippoc., amygdala   excitatory        gluta
make specific connections

Stellate
  inhibitory aspiny
  excitatory spiny

Basket                                  inhibitory        GABA

Spindle
```

Many neurons tend to fire in bursts, "making words out of the letters (spikes)," a higher-level symbol. This is likely to be most of the time just to stand out as manifest symbols against the noise rather than single bits, which supports resonance and non-resonance (not being too sensitive). It is also a genetically tunable intensity.

The cortex exhibits neural oscillations at these frequencies, which can be picked up on an EEG:

delta 1-4 Hz

theta 4-8 Hz   Linked to memory, induction of LTP.

alpha 7.5-12.5 Hz

beta 13-20 Hz

gamma 30-70 Hz   About 20 ms. Cognitive processing.

Delta frequency matches layers II and III of the cortex.

Alpha and beta match layer V.

Resonance occurs when positive and negative regenerative paths have stabilized into a perpetuating pattern. In this way pattern emerges, and can drive the slower building of further stable dynamic connections. This loop could also drive continued plucking of the joy field (which would be even more effective were there to exist there time summation in addition to spatial summation).

Resonant cortical columns form when stimulated at gamma frequencies. Their isolation is from active inhibition.

Brain modes

```
  Awake        EEG gamma 20-50 Hz, 50-20 ms
  REM sleep    EEG gamma 20-50 Hz, 50-20 ms
```

```
Non-REM sleep EEG <1Hz - 7Hz, Hippocampus
involved
    Stage N1: alpha to theta
    Stage N2
    Stage N3 (deep sleep): presence of delta
```

In deep sleep, TC neurons oscillate at 1 Hz and cause cortical neurons to alternately shut down and then fire a great deal.

The action potential is a traveling circular ring of charge and charge differential. The diameter of the ring varies by neuron, as does the length of travel, and there may be myelin present or not, but the diameter of an axon, fascinatingly, is constant. For interneurons the spikes won't travel much because the axons are short (and in all different directions). The interneurons are new to cortex.

More likely, the pyramidal neurons are the conscious ones. Layers II, III, IV, V would then be the conscious layers of cortex, especially layers III and V, and I and VI would not be. (Drugs that shut down whole populations of specific neuron types would be useful in research.)

Medically, what is needed for consciousness is the RAS and the cortex, and the RAS is made of very old tissue (brainstem) and serves simply to arouse the cortex. So, cortex is where consciousness is made, both macro and micro.

Note that you are not conscious of primary sensory processing; these brain areas are not themselves conscious (unless as small isolated blips). Only certain regions of cortex are conscious (though a great deal of it probably). It is also possible that sizable regions of the cortex are conscious and isolated, unfelt by the main You, like each brain half of a split-brain patient. The connectedness might dynamically vary, as islands fusing and splitting.

The cortex is mostly passive and ready to be stimulated in different ways. It contains about 80% excitatory and 20% inhibitory neurons.

Pyramidal cells are the primary excitation units of the prefrontal (and all) cortex. They appear especially in cortex layers III and V. They use glutamate and GABA. Glutamate excites them, GABA inhibits them. Both their dendrites and their axons are highly branched. Each receives about 30,000 excitatory and 1,700 inhibitory inputs.

The RAS (ascending reticular activating system) in the brainstem projects mainly to *prefrontal* cortex. The RAS provides the stimulation that gets the whole thalamus-cortex complex going.

The upper cortex across the top and sides of the head is cognitive. The lower cortex along the bottom of the brain case is emotional. We might also ask what makes paleomammalian cortex different from neo-mammalian cortex? Mammals are all quite emotional, but higher mammals add a great deal of information processing ability. This suggests that the new parts of the human brain are where not to look.

Once reptiles were in place, big emotional elaboration occurred with the mammals. When mammalian ancestors of the primates went up into the trees, there began major elaboration of visual processing and dexterity (which are cognitive).

Neocortex exists only in mammals, who make extensive use of emotionality, remember relevant things well, and solve more complex problems effectively than their predecessors.

The cingulate gyrus of the cortex, specifically the dorsal part of the ACC, is emotional. This has been found to be where pain (dysphoria) happens, *for both physical and emotional pain*. It has a lack of granule cells and is very unusual in having spindle cells. Spindle cells, which connect widely separated areas of the brain, only exist in areas associated with consciousness and emotion. But also, they exist only in the very most intelligent animals, whereas lesser but very conscious animals do not have them. Still, it is possible that spindle cells are either highly feeling cells, or useful in studies by being connected to areas that are highly feeling.

Basket cells are inhibitory and appear in areas like the hippocampus and cerebellum that are not associated with emotionality. The inhibitory interneurons of neocortex appear to keep the pyramidals and/or chandelier cells from getting too excited, unlike in older cortex, so that broader analysis and synthesis may percolate before a conclusion and specific action.

Pyramidal cells are very present in emotional areas, including the amygdala. Chandelier cells are intriguing. These synapse exclusively to the axon initial segment of pyramidal neurons, and very potently excite them to fire. They are more present in higher mammals. They are cortical interneurons.

It appears that some of the pyramidal cells, and/or the associated chandelier cells, are the conscious cells.

The right cerebral hemisphere is associated with negative emotion and fear (it thus might be interesting to study correlations of depression and anxiety versus math ability).

In primates, social and emotional "processing" (the consideration of these factors for making decisions) is localized to the orbitofrontal cortex. Notably, the OFC receives a lot of dopamine.

It does not seem that prefrontal recurrence with the limbic areas is necessary to enable the latter's capacity for emotion.

9.11. Pain and Pleasure Areas

We know the pain and pleasure centers of the brain. Pain and pleasure are fundamental consciousness.

9.11.1. Pain

Pain *signals* go to

- Somatosensory cortex
- Periaqueductal grey matter in the midbrain of the brainstem (subcortical)
- Insular cortex
- And in particular, the anterior cingulate cortex (ACC), Brodmann 24, 32, 33, where occurs the unpleasantness of pain (fundamental consciousness).

The neurotransmitter serotonin is associated with emotion and mood; bad levels in either direction may cause depression. (LSD, which causes massive malfunction of both the cognitive and the emotional, operates at these receptors.) Serotonergic neurons may be closely associated with the physical process for consciousness that we are looking for (perhaps by being one hop away from the neurons directly responsible).

The reticular formation (RAS) projects everywhere in the cortex. Interestingly, low brain activation occurs with depression, as though the brain is naturally depressed emotionally and must be stimulated up out of that.

9.11.2. Pleasure

In the case of pleasure centers, listed below, there is some question as to how much of this is cortical given our knowledge today; we can always use pain instead if need be to search for the consciousness mechanism.

- Nucleus accumbens of basal ganglia (subcortical). This has 95% GABAergic medium spiny neurons with D-1 or D-2 receptors
- Ventral pallidum of basal ganglia (subcortical)
- Parabrachial nucleus (subcortical)
- probably orbitofrontal cortex
- probably insular cortex

The olfactory tubercle plays a role in transmitting positive signals to reward sensors.

Reward appears to involve a cortico-basal ganglia-thalamo-cortical loop. Reward appears to involve glutamate interneurons, GABA medium spiny neurons, and dopamine projection neurons. The neurotransmitter dopamine is considered a global reward signal (can be either excitatory or inhibitory at synapses). Neurons in dopamine pathways are prime candidates for the physical interaction we are looking for.

9.11.3. Both Pain and Pleasure

Anterior cingulate cortex is implicated in both reward *anticipation* and the unpleasantness of pain. It is neocortex. Crick thought this the center of so-called free will, and it is one good candidate. The anterior cingulate cortex is unique in its abundance of spindle cells (recent in evolution and found only in very smart mammals). The OFC also participates in the *expectation* of actual reward or punishment in response to stimuli.

9.12. Evolution

Pain and pleasure have been discovered and harvested by evolution. There are likely other worlds in which pain and pleasure don't exist to organisms, even complicated ones.

Pain and pleasure might not actually be the same force. And there might be animals that feel pain but no pleasure. Use of one probably came first in bio evolution (and that would likely be pain).

Intensity of experience increases by more spots, and indirectly, by more related interconnections.

Harvesting fundamental consciousness by evolution would mean "crafting" the shapes and/or construction and constituents of neurons, and possibly sets of neurons.

We can recap the evolutionary development sequence because the needs and difficulties are obvious, and because we have the animals around us as evidence.

First needed would be basic stimulus/response. This could be viewed as the degenerate case of the cognitive. Then would come totally dispassionate ability for slightly complicated perception and actions. Calculating ability- pure, passive cognitive- would come first, and exists in primitive animals.

Feeling provides for the first operant learning opportunity, and certainly for more complicated organisms. It offers a broad whatever-you're-doing-now-is-bad (or good) signal for learning, and provides the second needed broadly distributed system- Value- which determines what action impulses are let through, the other being Activation. Third is Attention, targeting and focusing the cognitive portions. (Even before well-developed learning appeared, feeling might allow for generalized *real-time* changes in decisions in creatures with some neural complexity, and serve Attention). (Some behaviors may appear to be operant learning but not actually rely on feeling. For example, an electric shock might cause some disruption or condition that is not pain.)

This progression would have begun in the evolutionary outgrowth of the allocortex- the olfactory system and the hippocampus. The paleocortical piriform cortex a.k.a. posterior orbitofrontal cortex (posterior OFC), is heavily correlated with the cingulate gyrus and the septal area, which is part of the basal forebrain. The OFC mediates the expectation of reward/punishment in response to stimuli. The paleocortical olfactory tubercle plays a role in transmitting positive signals to reward sensors.

Once all these lower mechanisms were in place, they could be extended (scaled) greatly and the neocortex burst forth into high growth. Warmer/Colder is a very powerful concept (complementing Recognition).

PART II:
HIGHER CONSCIOUSNESS

In this Part of the book we consider higher consciousness (that is, higher than Fundamental Consciousness).

There are various things from which we could (and do) fashion a sort of *brain*. But consciousness allows what we call *minds*, which feel (and which find themselves to be terribly important, because they are so relevant to themselves and each other).

The difficulty in trying to understand is not dispensed with by saying "let's not talk about consciousness, let's talk about feeling." Because at the human consciousness level, there is still a party doing the "feeling" to contend with. The brain is a bottle containing separate fireflies of feeling, but they are connected by cognitive links, and possibly links which themselves feel.

A "feeling" at our scale is caused by a network of neurons co-stimulating each other and causing base feelings that add and fuse in the joy field. That is Me (at rest, unthinking). If one neuron stops I feel no difference. If they all stop, I stop (even if one remains active).

Base feeling occurs and fuses in the joy field; advanced emotions have cognitive content. Cognitive processing and knowledge are carried by network structure. In principal, some sets of neurons could participate in both. The author suspects specialization, with different areas of different evolutionary ages doing mainly one or the other.

There is also an emotional "logic" to higher feeling that mirrors that of the cognitive. Whereas in cognitive areas, rivulets of recognition compete and build into a recognition conclusion, in the emotional areas, rivulets of neural flows build to a conclusion of the overall positive or negative value of a recognized thing. For the latter- the "emotional processing"- the neurons involved may be the same ones that feel, or may be sister networks embedded in the composite emotional neural tissue (cortex). The emotional conclusions will drive what to pay attention to, and which thing to do.

Regardless of the brain's use of joy field summation, the cognitive *circuitry* is necessary for conscious experience. While there is no indication or necessity for quite such precise machinery in contact with the joy field, cognition happens in our refined neural networks.

What we humans commonly call consciousness and awareness amount to feeling about our thinking and thinking about our feeling, in a tight back-and-forth.

The full picture of the pervasive low-level consciousness, is of reactively feeling our thoughts, cognitively noticing our feelings, back and forth. It is the interwoven, co-stimulating mesh of the cognitive and emotional. Atop and co-interacting with this is another layer containing language. The lower consciousness reacts to phrases, especially the internally generated ones, while the higher, language portion reacts to what it is observing with phrases that summarize and organize, and around it goes. This triad is the architectural spine of mind.

We think and feel together a great deal. And as we cope and mature, we construct emotional steadiness around child minds that are highly feeling. This development is elaborate. First come immediate impulses from raw higher feeling. Then appears feeling-driven deliberative mental action, which we call "will." From there, we advance to willfully controlling our emotional brain areas, which we call "willpower."

It is the physical body that provides the organization of a simple force- in fact, several. Otherwise there would be no mind and no intelligence.

Life began without sentience, without feeling. The very first neurons probably did not feel (or not very clearly) either, as there was no need and we have many neurons in our head that probably don't feel much. Life discovered and organized sentience. Thus, we surmise that very primitive animals are not conscious.

Just as the "consciousness" and behavior of a society is emergent from those of individual people, the "mind" is the behavior of a population of small "voices." For even small mental things to emerge from background noise, several constituents are generally needed, and they must cohere. On a larger mind scale, emotions provide an out-of-band cohesion in space and time, a summarizing score as to how things in general have been going, and whether large changes in choices are necessary.

Note also that some brain regions could contain local regions of sophisticated consciousness that do not reach *the* fully-developed and unified consciousness that we call our mind. Split-brain patients illustrate this possibility (as do all other people, whom we assume are conscious).

10. From Association to Thinking

Thinking is mental action- studying, as though with our eyes, sets of objects that appear in the mind by sub-mind processes, and manipulating them, as though with our hands. Actions of all kinds are directed by will, which emerges from feelings. The ability to then execute a series of action objects underpins language. Language allows for more precise and useful organization and more precise and longer serialization. With this running commentary to ourselves we convince and persuade ourselves (or more precisely, some portions of self convince and persuade other portions).

Like lower feeling, association is spontaneous and undisciplined. When motor precision is brought to bear, thinking becomes a real and elaborate act of its own, of a machine that can be relied on in complex situations. The total mind is a mix of free-run and controlled mental movements. The free-run is generally "sub-conscious" and ineffable. It injects things into our true mind, objects from some areas and words from others. It is our mind's eye and our mind's ear, which recurrently see their results and respond with related, new such. Meanwhile we manipulate- move and modify- these objects and words with our motor- and speech-related faculties, and the free-run portions react to the new patterns so made.

To our minds, almost everything is either an object in space that we manipulate, an emotion, or an utterance. These concepts are generalized to represent nearly everything we think about. This presents an opportunity for humanity to construct a "physics of mind," meaning not the underlying physics in this case, but a correct theory of operation of this macroscopic world of nonextended things.

We are used to observing things from outside them. In consciousness we are experiencing from the inside of the physical process, actually we are living it, actually it is just happening, while simultaneously being watched by the cognitive half of Me.

11. What It is Like to Be a Human

Higher consciousness experiences time, as a consequence of the differences it sees between its world at t and t0 (where t and t0 both keep increasing). Effectively, consciousness moves along time. At the low level there are linked stimulations of conscious events, and the propagation times and stimulus persistence of the brain. These cause the continuous flow of High consciousness time and events.

It may also be said that consciousness *causes* time, in the sense of the experience of a physical dimension that would otherwise just be another physical dimension. One might say that time is the progression toward increased entropy, or one might say that time is the *experience* of progression toward increased entropy; this contrast is just a matter of definition. The idea that the consciousness of beings such as ourselves literally causes physical time, however, we reject. There is insufficient evidence to accept that possibility. It remains possible that the joy field and time have a close relationship.

Time might be a sequence of quantum collapse events, but space might, also.

Humans are action-oriented. The brain exists to decide what to *do*. The frontal lobe is the motor lobe. Impulses for action are suppressed and analyzed and selected and organized for submission to execution areas. Unlike other animals that go to sleep whenever they can, humans are curious and exploratory and chatty and get bored, stir-crazy, and lonesome. This keeps us active and learning.

The human brain features high-level, complex connections: understanding is pleasurable, novelty is pleasurable. These make curiosity and enhance creativity, itself a mix of compliance and variance.

Things bubble up into human consciousness from "subconscious" areas. Only some parts of the brain are organized to be part of the human-conscious. It is the unity of feeling and the unity of the cognitive and the fusion into a greater unity of these two (plus language) that form the experience of what we call our mind. For the base feeling this fusion is done in and by the joy field. For the cognitive, this is done by networking of neurons. The final fusion is done by the connection of cognitive (information-processing) neurons with feeling neurons, which connects cognition to feeling. This creates both complex deep emotions with a cognitive component, and rich thought embroidered with, encouraged by, discouraged by, and steered by emotion. The two things that make all this possible are the information processing and communication ability of all neurons, and the (strong) connection to the joy field of some neurons.

Organized mind appears when awake, as opposed to when dreaming. One possibility is that the mind would be organized when we sleep were it not for interference caused by signals that are generated (or removed) during sleep. Another possibility is that the operation of perceptual regions close to the senses, which are inoperative during sleep, ordinarily organize the flow through the rest of the brain, to restrict what is thought to what "makes sense." Both factors could well be significantly contributing.

To "make sense" is a pleasurable condition related to high discovered recognition and correlation, and low ambiguity, as seen in strong standing signals against a relatively quiet background. It brings comfort and is sought. Confusion and unsureness are negative, and we seek to eliminate or avoid these. The neural circuits for this are imaginable.

The human mind loves order in all things. It has grown through evolution to where it is possible for it to see a whole big picture (and even to describe that to another), wherefrom flows its great power.

Note also that *emotions* are quite in their normal form and understandable during sleep; it is the cognitive, which is intrinsically of structured nature and of high continuous relatedness to many patterns we know, that loses good form and becomes disjointed during sleep.

A possibility is that our simulation centers- our imaginations- are only constructed to run coherently for seconds at a time, because that's what is needed. In sleep we also don't playback our memories verbatim in long sequences; memory centers are executing a consolidation process and bubbling and burping as a result when they are active at all. Thus, disconnected splotches of memories start up short sequences of predictive simulations with no sensory perception or memory circuits activation with which to police for order.

Human-level consciousness occurs from adding to lower animals

   a) Full language
   b) A strong sense of what "makes sense"

(b) is what is not acting as much when we are dreaming. We may say that a scene or utterance is "incomprehensible," which means it does not comprehensively hang together (make connections) that result mostly in recognitions of things and relationships deeply honed over the years, rather than confusion resulting from a lack of this.

Things "make sense" when there is high recognition and sharp inhibition, connected to feeling. In sleep the contrast is lost, and things bleed over into each other.

Humans are so advanced that they are self-aware of concepts as high as feeling confused and feeling sure they have "got it" (which are feelings and may even be misplaced ones). Knowledge is not feeling. But feeling convinced of knowing something is a feeling (and generally a positive one). "I feel that X" means that X is coupled with a feeling of some certainty.

The mechanics of a working mind run on Recognition and Value. Direct pain and pleasure served as the first source of Value, allowing bootstrapping. Now, many local inter-neuronal circuits have likely generalized Value informationally, and no longer feel pain and pleasure, there being no evolutionary need to maintain them. At the micro level of large networks, the original pain and pleasure are too crude- at least, globalized pain and pleasure are too crude.

Mouse consciousness is emotional and very action-oriented. It lives in a world of smells and touch and motor movement. In contrast, a grown chimpanzee has the intelligence of a 2.5-year-old human, and probably his emotional life as well. Such a human is no longer an infant but a toddler, with rudimentary language ability.

Human consciousness is visual and analytic. Selective attention is only needed in a complex cognitive world, and humans inhibit action until selected and initiated after analysis of the world.

Consciousness has duration, necessarily. The actual "remembered present," as Edelman called it, gives a more *global* (across space) and *constant* (across time) existence to an object than the millisecond/millimeter world of neural-neighborhood-scale operations.

This very short-term memory is very useful to sophisticated cognition, a sort of object cohesion and permanence enabler. Coupled with the connections that compare between different parts of this time snippet and those that react to it and supply it with emotion, to a first approximation this *is* low-level, higher-animal consciousness (that which makes a sentient being).

Consciousness has a smidgeon of usefulness even without net emotional affect. Change and novelty and distinct form cause this impact (and on repetition it fades, both as it happens and from (nonmotor) memory). But consciousness without net emotional affect, which supplies positive and negative values, is very limited in what it can do.

The emotions we feel affix value to things, including objects to study and courses of action from which to choose. In principle another mechanism could have been used to supply Value, but emotional feeling is what evolution discovered in Nature that could be used.

And why did evolution use this means? What is helpful about feeling, is that it has its own intrinsic value- positive or negative. That means evolution did not have to manage to use valueless connections in a consistent way to emulate the same thing, or invent centers as assigned positive and negative places to connect to. Consistently labeled positive and negative things were available in nature to happen upon, and then use again and again. This made it easier, and thus more likely, and thus the way, that summary value is implemented in the brain.

We don't call it thinking unless we can feel it happening. There are also underlying processes to our thinking which we can't see (associating). Crossing the boundary is what happens when things "pop into our mind." We are able to deliberately reflect to encourage these unconscious processes to happen, and to deliberately turn down our attention and fixations to allow broader and freer flow of proto-thoughts and metathoughts.

We can feel the thinking process partly because cogitating kicks feeling parts (rather like feeling limb motion), and because of the changes in blood vessels (which we can see in imaging).

It is probably even important to our thought processes that we can feel our thoughts and where we are paying attention, and probably there is specific sensation of that because the sense-ability has utility.

General "awareness" may include arousal (which increases intensity of attention behavior) more than emotional response, which gives value to things. A high level of arousal can be pleasurable or stressful; it is an intensifier.

Our emotions are different. We can just feel sad or fearful, *then* start thinking or noticing or inventing supposedly why. Sad can be purely dysphoric, without knowing "why," which is cognitive (in contrast, the more complex *dread* is also dysphoric, but *refers* to something).

People use the word "feeling" to refer both to emotions and to intuitional judgements. This is natural, because our processing circuitry includes a bed of elaborate, general recognition circuits effectively much-interwoven with feeling circuits so as to develop finely-detailed value processing of things under consideration, resulting in a net high-level score, and the distinction between feeling and cool judgement can be difficult to tease apart.

It is possible both to have a "feeling" (an observation) of a large object in your face without net caring, and also possible to feel positive or negative for no apparent reason. As to the former- impact- it is the shapeless glob of emotional feeling that is doing the reacting to factual stimuli. You still hear emotional white noise even though it doesn't settle on organized tones. The white noise activity is impact. Everything makes "you" feel something- even if only a unified agitation.

Concentrating of pain or pleasure is relatively unusual in the universe (rather like the very big molecular machinery of life being unusual). Thus, it is a specific achievement of evolution that is useful. Pain and pleasure allow advanced minds. By the Anthropic Principle, that is why it's here.

Physical consciousness was exploited in vertebrates just as the laws of mechanics and chemistry were. It serves the functions of learning and assigning general goodness and badness to large and complicated and long-lasting experiences and plans, and in bonding social groups. Emotions support this in mammals, and the values system that make one mammal mind so powerful.

An emotion of some mammals is loneliness, one of the most advanced, in fact. (And when people get together and become so close they don't have to speak to communicate, they can in effect form a larger mind, such as a couple.)

Higher consciousness is like multicellularity, something that emerged as a complex system incorporating many points of occurrence of base consciousness, allowing novel survival niches by supporting highly elaborate, flexible, and effective structures- in this case, our brains.

Our consciousness is observable and organized. This does not mean simple consciousness does not happen all around us. The organization of fundamental consciousness into high-level consciousness was done to our brains by evolution (or preceding mind to whom we are the "artificial" intelligence). As one of the available forces, fundamental consciousness probably has been exploited elsewhere, and elsewhere there will be pain and pleasure (and therefore perhaps good and evil).

Is consciousness "just" an epiphenomenon? Yes, in the sense that it was not necessary that we actually feel, only that the same values be attached to things as feelings allow. No, in that the values consciousness provides form a key part of what our minds are and how they work.

Value is needed for how we think. It is what steers the cognitive centers to make decisions. The cognitive centers can recognize things and recall candidate action sequences, but it is the emotional input that determines the selections- the "voluntary" behavior. Volition equals emotion.

I feel. What is I? An assemblage of interlinked cognitive and feeling events. I am intellectually aware that I am feeling, and I can feel and have feelings about my thinking. The human mind is everything it is because it contains not only thought and feeling but can think about feeling and feel about thinking, and do so under the organizing and evocative presence of language. Higher consciousness is the interaction of knowing and feeling, and each of each other. The deep mutual interplay of feeling and thought, and of these with language, comprise the mind.

12. Language

Language organizes thought; most complicated non-primates are disorganized and react to mere fragments of what of we would call real understanding.

Language organizes all the little fits of proto thought, and one's own assembled phrases stimulate those same fitful centers to recognize and remember and churn out new proto thoughts. Language crisply organizes associations and spatial, motor, and intuitive thought. This is done using the language centers of our brain. The appearance ("hearing") of the sentences to our non-language brain centers evokes in them the unstructured thought that is then organized into language again. This cycle, along with emotion, is the fundamental stuff of our highest-level consciousness (mind), as we talk to ourselves about our thoughts and feelings and react to what we have heard from ourselves.

We have many words for various kinds of feeling (and thinking), which attest to the richness and complexity of human-level minds.

13. Brain Structures for Higher Consciousness

The richly emotional sections of the brain, (literally) closer to the ground, are both separated from and connected to the (literally) higher and stoic, calculating portions of the brain.

The circuitry of thalamus and cortex form recurrent neural loops.

Layer IV of cortex takes the input for all senses except smell, and these are received from the isothalamus portion of the thalamus. The cerebellum is also treated in this way as an input organ. The inputs are sent to the prefrontal motor area as well as the sensory areas.

The TRN part of the thalamus receives cortical signals from cortical layer VI and regulates activity in the thalamus. It takes much more input than it produces output.

See the discussion in Appendix IV for more.

13.1. Theory

In our theory, actual consciousness does not arise directly from resonant signal loops, but from the consistent stimulation of the joy field that such loops provide, by repeated operation of the basic stimulation action probably associated with neural firing. The specific resonant frequency is unimportant for this, it is whatever the brain loops happen to settle to.

The macro-sized consciousness is also thus probably not a quantum effect, even though the individual fundamental feelings operate at a quantum scale.

It is theorized (by others) that neural synchronization accounts for the neural basis of consciousness.

At a fundamental consciousness level, no, it doesn't. At a concentration of consciousness level, yes. At a much higher level involving attention, yes, by locking on to what is of interest. A competitively selected, regenerating loop provides a lock on the item of interest.

First, basic cognitive ability was evolutionarily achieved through the sequentially-performed neural triad of candidate recognition, candidate ranking, singular final selection. The first and last are easily done by relatively primitive means. The second is where the circuitry elaboration probably occurs in high-end brains, as this is where the scoring mechanism is applied in the competition among proto-ideas.

We know that the thalamus appears to gate, and thus determine what is paid attention to. It receives input from cortex, which is likely elaborate conclusions, in making those gating "decisions." The gating thalamus and the analyzing cortex cooperate, settling on what to settle on, which is what is found by cortex to be most interesting, which will be what produces the most related activity of inputs and memories. This (decision-making as to attention) is probably another cortical takeover of functionality that used to be done by the thalamus alone. This takeover then enabled intentional and deliberate acts of attention.

13.2. Conjectures

Conjecture 1: The cerebellum shapes the actions of the messy brain to bring neat order of execution to the intended acts started by its willful spasms. The cerebellum shapes thought motions- movements of the mind- just as with limbs and fingers, to allow for precise manipulations and basic connection and flow of action sequences. In this case the actions are on mental objects. Introspecting, one can perceive such actions as putting a (mental) item "off to the side," after getting a good "grasp" of it. It is known that perceiving a physical object is closely tied, circuit-wise, to manipulating it. Mental actions require some coordination, too.

Conjecture 2: Cortical columns/modules may be exhibit a highly independent consciousness, co-influenceable like the two hemispheres of the brain. Also, some of them may be more dominant than others, to promote organized action.

This would mean a physical hierarchy of levels begetting consciousness, from neurons, to columns, to interconnected areas composed of columns.

14. Attention, Will, Willful Attention, and Attention to Will

To care is to have emotion. To care what happens and try to affect that, is will. The self-happening nature of feeling, interacting with the cognitive, results in "my" deciding based on my own "free" will, which amounts to emotion determining what you do.

Yes, feeling is the one thing that makes us different from a machine, it creates what we call "free" will. But genuine feeling and genuine freedom are not the same. Will free from what? From the nature of other things and persons? Yes. From its own nature? No, it is a product of the brain's structure.

The brain has evolved to the point where it can control even itself and feel itself thinking. Many actions and inputs are internal to the brain. The highest organisms are not only aware of themselves but aware of their ongoing thinking. Control is exercised by active attention.

The importance attributed to cognitive objects- by emotion or novelty- appears to be cortical.

The cerebral cortex forms a network wherein almost all its (nonolfactory) input comes through the thalamus, with returning fibers going back to the thalamus, such that mapping is precisely maintained between cortical patches and thalamic regions. This means the brain can modulate what input gets through based both on unconscious tracking of conspicuous things like motion and loud noise, and on desires. The gating through the thalamus constitutes the final act of attention, while the cortical input to it designates what is desirable to attend to.

Note that given emotional cortical processing, people to varying degrees can desensitize to the emotional activity volitionally. This amounts to a battle of emotions, since will itself is emotional. (The degree of influence of will over immediate emotions indicates personal ability to run "cool" rather than "hot.") We are also able to cognitively watch the battle itself occur, and react emotionally to it.

See the appendices for more on this chapter's topic.

15. Self and the Unity of "Me"

Evolutionarily, impulses precede a single, coherent self. Unity of mind as an acting agent precedes any awareness of unitary self.

The unity of an experience of ours, which is more elaborate than fundamental consciousness, involves:

- Adding and fusion in the joy field

- Co-stimulating of the field from multiple brain points

- Neural links, which may themselves stimulate the joy field

- Cognitive observation of the feelings occurring

In addition to a fused emotional experience, we *notice* a commonality of feeling; the noticing is a cognitive thing.

A skeptic to the arguments we have made might say a possibility is that the "feeling" of being one is a completely cognitive process- a noticing of a simultaneous band of feeling points. Feeling points can't notice each other, a cognitive part would have to notice them all. The argument would be that an emergent feeling of coherent self seems like one feeling building because each little bit of watching and deciding in the brain is getting its own feeling input, though each bit of that feeling is alike.

At first it may seem that we don't really know the feeling adds and fuses- that it simply *seems* that way. But imagine getting more and more excited in a pleasurable experience. You are *feeling* more *intensely*. What you are feeling you are feeling more of- it adds together. It is not a happening of more points in isolation, and it is not something you just observe more of. The experience is not just of an increase in number but an increase of feeling intensity.

Even if sentons can have different energies as can photons, a more stimulated neuron or group of neurons does not fire with (much) higher amplitude (intensity), it fires with higher frequency. This means space and/or time summation must be at work.

It has been theorized that thalamocortical oscillations from recurrent pathways between different brain areas gives the sense of unity of mind. This might work for the cognitive (and such a process is probably at least as much related to having many common associations firing and supporting each other and thereby inhibiting less well-known and more confusing currents), but it does not explain the unity and amplitude variation of feeling.

So, let us now examine the sense of unity of mind.

Imagine two or more people together, each feeling, and seeing each other feel, and personally reacting to each other, so that it is a shared experience. In such a situation as that, you don't actually share the same exact happening feeling, you experience another instance of feeling similar to the feeling the other person is experiencing, which is evoked in you because you are seeing the same scene and possibly watching each other having similar reactions.

Now, internally, "you" is a family that knows each other well and experiences and feels for each other. This connected and coordinated happening is the unity you experience. The feeling part of you has this commonality in the joy field, not just as discrete emotional charges. It is very intimate, it is real oneness.

Additional cohesiveness comes in at a higher level from the vast neural network *knowing* things, and then comes your higher-level "feeling" that you do know those things (a feeling of certainty, which is a composite incorporating positive feeling associated with a cognitive conclusion). There is also a cohesion of knowing and cognitive attention (the attention being driven by cognitive novelty or by feeling).

Conversely, "I" can also feel confused, especially before letting things settle for a few seconds, and have multiple feelings at the same time, so there is not a single global net feeling. "I" feel scattered. The settling down is the process of some flows increasing, finding each other, and joining in positive regeneration, while others are inhibited and peter out (from what electrical engineers call Automatic Gain Control).

When we say we are having "mixed feelings," we usually mean (our attention is) flitting from one feeling to another at sequential instants, and thus have multiple feelings during a slightly longer "same time". We imagine it as with a butterfly flitting from place to place (per James), because of the such-like captures of our attention, but it is actually swells of various places of the emotional sea embodied in neural tissue.

When many (feeling) neurons light up, I may say that "*I feel* happy" because that is what my cognitive mind can understand. But what is happening is that joy is itself happening, located inside my head. At the same time, multitudes of neurons observe the same stimulus and each other and the simultaneous joy.

If some neurons die, your emotional and cognitive energies diminish, but even when many specific points of contribution have been eliminated the experience of unified feeling remains. It is more an experience like one with little currents in it than a handful of little light bulbs or other hard, separate objects that are on or off. Taking another cue from the rich source of established choices of words, we say we are "in" a good mood or "in" love. This lake of feeling is Me. Your self is not as much an object, as a cloud.

The atoms of feeling are bound together and affect each other in two ways: cyclic regenerative pathways at a higher level, and addition and fusion in the joy field at the fundamental level. In the first case, each bit of feeling happens by itself, and in the second, one charge cloud stands united by virtue of locality and the laws of (complete) physics.

The initial reason for a human feeling to be strong, is that the initial cognitive channels of perception or recall synapse on many feeling neurons. Once so ignited, this group may regeneratively amplify itself for a period.

The singleness of self you internally *observe* may come purely from the cognitive connections. In many places in the brain these cognitive connections bring separate things together into a conclusive whole.

As a mind, *you* don't feel *something*, you *are* your feelings. You are the cloud of feeling your brain makes. You also have a cognitive processor and these two things are heavily interconnected/intertwined. What makes you very special, though, is not that you can calculate, as many things effectively do, but that you can feel. In making that so, evolution has caused you to value some things more than others. Ancestors with different values did not survive. ("Value" is an interesting word because it allows us to discuss cognitively that which is aesthetic.)

Realizing you are a cloud or a lake solves both the Observer Problem and the Unity Problem. And at any time, the unified you can lose a few cells and lose a little bit of feeling and you are still there, because you are not a single point but a cloud of these touching points.

"Feeling" "by" "Me" is a larger event than one senton (and for that matter a single photon is generally imperceptible to us, too). The feelings our selves experience are each thousands or millions of these tiny spots of emotional light.

When there is an accumulation of many matter particles in one place, they will pull hard with the force of gravity. So, too, many joy-charged particles together will pull hard- will create a large charge field. That charge (or its movement) causes feeling- *is* feeling.

There is a perceived sense of self. And there is a cloud of feeling. We have both, and they interact. The mind is this thinking + feeling. The emotion cloud travels with our thinking brain.

Cognitive unity does not seem much of a mystery as to how it is possible and how it could be made: connected neural processing networks. We can see (the removal of) such unity at work in split-brain epilepsy patients.

A sense of self could start with bodily self- the spatial- which for us is smoothly cohesive and continuous.   The sense of self is a confluence of integrated body sense, self-talk, and will, which is emotion. Some say that the insular cortex (which is emotional) gives the sense of self. The seat of sense of self has been located, some say, in anterior cingulate cortex. In any case, the prefrontal cortex does the high thinking, steered by the emotional cortex on the cerebrum's lower surfaces, with calculating and perceiving happening in its upper surfaces.

At your very deepest conscious core "you" are your emotional feelings. Intelligence, which is cognitive, is poured atop that, making you not only able to feel but to reflect and imagine. Language is poured atop that and thinking becomes organized and step-by-step.

You are attached to the consciousness field, the joy field. "You" *are* your present spot in the consciousness field. Within it, each feeling point can feel others nearby, in the sense that each influences each (field summation), to be in the same direction as itself.

The feeling is additive, the charge sums. The joy quantum field itself does this is as you perturb it at multiple points. You are the place in the joy field that is being perturbed by your brain.

Pain and pleasure are real. You don't feel *that* you are having pain, you are *in* pain. Mood is probably semi-widely distributed, while attention lights up the specific little brain regions of interest. All of this collides in the prefrontal cortex, and choices result, as a system-level effect.

16. Destiny

There is indeed a breath of life- an animation. That breath is (organized, concentrated) pain and pleasure, it is the soul.

This thing- the soul- is indeed different and apart from the matter we know, and real. But on death it dissipates as does all the other matter and energy, and in the same manner, returns to the background, a background of dimly feeling dust.

You are a child of the Universe, and you will return to this nameless dust. The Universe itself will at least grow cold if not die, so even if you lived up to that point, it would make no difference beyond it.

Will is nothing more than what you want to do, overall, and that is whatever causes the most pleasure and the least pain, overall. This is why we train children to be pained by doing things that are designated wrong, and has as inputs mammalian traits of empathy and kinship. And we can train ourselves.

There is no practical conflict between "free will" and destiny (though for some reason many have anguished over it). Like everything else in Nature, you perform to your nature. You say you are free to do what you want. What you are definitely not free to do is what you don't want to do, all things considered. But, by definition, that is not desirable anyway.

You are a mechanism, but one that can experience pleasure, and can (and shall!) seek more of it. And pleasure is as fundamentally real as light and matter.

17. Why Consciousness?

This question applies to consciousness at all its levels.

Why this way? There is no Why this exact way, it is one possibility.

Higher consciousness evolved. But fundamental consciousness did not evolve (except perhaps during the earliest physics of the emerging Universe).

Why is there consciousness at all? Based on the Anthropic Principle, there is consciousness in our universe because that is a way that intelligent beings can exist, to then ask that question. This also answers why there is suffering (and pleasure) in the world, which amounts to the same kind of question.

The question of why, then, are there such beings at all is not really different from the question of why there is anything else. The answer is that it is possible and there is no reason for there not to be, so there can be, and there happen to be.

Everything needing a reason is a human notion, based on the goal-directed and problem-solving urges (pains and pleasures) and behaviors we have evolved, having discovered macroscopic causality, so that our genes would survive rather than others'.

There are probably other "places" that also exist as much as ours does, where much of what we have here does not and cannot exist. For nothing to exist anywhere at all, would be a very special and unlikely case! A preference for that is a quirky human prejudice.

We know that it is possible for things to exist because things do exist. There is no ultimate base Why, or any fundamental requirement for one. Whereas humans are possible, and so might exist, and happen to, many other imaginable things happen to not exist.

The answer to "Why is there consciousness?" is:

"Why" is a human concept and need. Consciousness is possible and serves the evolution of creatures that can ask why, and thus we find ourselves in a Universe having everything necessary for consciousness to be possible.

18. Artificial Consciousness and Trans-Human Fused Consciousness

Real artificial consciousness, entirely different from AI and not needed for truly-intelligent AI given simulated consciousness, is in principle doable! By this we mean creating truly conscious minds that are made by machines we build rather than biological brains. This means that their pain and pleasure, unlike simulated consciousness, would be *real*. They would be like us (and thus perhaps entitled to rights).

Simulated consciousness is like a simulated car or solar system. It would mimic the same behaviors, but would not be real, not even a new form of real consciousness. That is because our minds are not just information flows, they have feelings. And feelings are real just as much as the steel of cars is real, they are just a different physical thing than moving steel.

A machine of today is not really conscious- not sentient- but can be effectively conscious-like, like a heat-seeking missile or drone. If we discover how to manipulate the joy field with an apparatus, then we may be able to make machines that are truly conscious. The specific electrical pattern of the traveling wave in a neuron's axon is different from any other pattern encountered in nature. This pattern could be produced artificially, possibly resulting in not just simulated feeling but *actual* feeling.

Artificial emotion could be achieved by creating engineered structures (perhaps nanostructures) similar to the critical elements of the brain. And there is a potential practical application for this, in augmenting and continuing our brains, not with simulated unreal emotion, but with the real emotion that fundamentally and satisfyingly makes us us, rather than models of us.

How could an experimenter tell the difference? In a practical way. Method acting is the easier and more convincing technique because the feelings you in the audience are observing, with their myriad cues, are real. Trying to control every muscle movement in simulation like a Laurence Olivier is very difficult for even a real human to pull off convincingly. To our advanced brains, not-truly-feeling minds are likely to be detected or suspected, even though in principle they might fool an observer. Advanced and consistent emotional mindfulness exhibited by a myriad of behaviors and utterances with all their attendant details and choices will probably be detected correctly as real. This is a spin on the Turing Test, where rather than purposely being remote from the new actor, you are purposely able to observe closely.

Providing the random numbers needed to support the mind algorithm by real random events in our universe (which ultimately come from quantum waveform collapses), rather than pseudorandom calculated numbers, would complete the entry of this consciousness into a full joining with our real world. It would no longer in any way be a simulation, but rather, a constructed actual conscious mind.

Note that genuine conscious feeling is not required for intelligence- a nonfeeling stand-in that did exactly the same function would do (and would basically appear to feel, even to itself), but it would not feel, it would at most think it feels. It wouldn't actually know "what it is like," which means to recall similar situations for the feelings they evoke. Its knowledge would be more like book knowledge. But genuine conscious feeling would not be necessary to display *behavior* equivalently "willful" to our own. There are those who believe that a simulation that, in interacting with us, is undetectable as artificial, must have true consciousness like us. There is no scientific basis for that belief whatever.

The first steps in trans-human fused consciousness would involve sensory, motor, cognitive processing, and perhaps simulated-consciousness implants. There would be new things to experience by the existing, re-purposed neurons, creating somewhat different new conscious experiences.

The next step would be joining true artificial consciousness with human consciousness. An advantage of this would be continuity as the original biological brain slowly dies off. In a way much more intimate than one's own child, the baton would be passed continuously from the original brain to the joined and fused and co-experiencing brain, to the surviving portion of brain. This is fundamentally different from a clone or a download, which ensures a legacy but not a survival, except of data patterns. Before the advent of fused true consciousness, you can only pass on what amounts to a book. After its advent, you instead change slowly, as in normal life, but your soul, in the same patch of the joy field, is poured slowly from one flask to mix with another, and then slowly evaporate to a degree, together. Though the atoms in your own body are continually replaced, you don't question whether you are still you after years of this. At a minimum, your most intimate imaginable "spouse," has lived on, indefinitely, and probably with better memory- your memory- than we all have today.

Until such time, we might work to preserve our feeling neurons as long as possible and replace the rest, which might make preserving what matters of your brain and surviving as a person, easier. How you feel and how you think are who you are, and the latter is only information processing. Learning somewhat different ways of adjusting our mechanical functions does not feel like a key change and loss of self, any more than does adopting an artificial limb.

On the journey, we may be able to feel entirely different forms of consciousness, depending on what we are attached to. We have foreseen such an experience in a clear way: lust is a conscious experience a small child does not have or comprehend, yet the same person later does. New-sense implants will come with new consciousness. They will cause new things to know "what it is like."

CONCLUSIONS

19. Conclusions

We conclude that there is a physical consciousness force, which we call the joy field, accounting for fundamental consciousness. Multiple pieces of evidence suggest this to us, and they are not all required; there is plenty of good reason to believe there is a joy field.

Evidentiary Element 1 is the tendency of Nature to repeat herself, using the same consistent patterns and methods for the phenomena of our world.

Element 2 is Occam's Razor: It is simpler to expect the same sort of mechanism at work than to presume a completely new one.

Element 3 is related to Element 1: In the human history of discovery, all phenomena have fallen to the same, increasingly generalized, scientific model.

Element 4: We can see that consciousness happens in the brain, and is almost certainly evoked by the brain, a physical object.

Element 5: To explain what supports yet smaller decompositions of experiencers and things being experienced, there must be a freestanding feeling.

It is probable that consciousness functions by the same rules (including mathematics) as all other phenomena that exist in Nature.

Animal-level consciousness is probably the result of activity of networks of physical neurons.

The nature of our human-level consciousness is that we react emotionally to (and perceive) our thinking, and we cognitively notice our feeling, and we talk to ourselves about all of this in a chorus of stream of consciousness. Our words then evoke feelings and more thoughts and around it goes. We talk to ourselves like a friend (or enemy), and our real friends' words are in there, incorporated.

APPENDICES

Appendix I: Reality, Existence, Universes

In trying to understand consciousness and base physics, one quickly runs up against the question "what is real?"

Reality is the totality of what is real. What is real?

A first stab at an answer is that something qualifies as real because it is made out of matter and/or energy, instead of being only imagined or observed or possible to be real. Real means to be made of something or to change something that changes something that changes something that is made of something- matter and/or energy. At base are particles, which real things are made of.

Descartes concluded the reality (and thus possibility) of himself, and thus the reality and possibility of existence, with "cogito ergo sum." (How did Descartes know he was thinking? Because he could sense it. Because he was conscious of it. Descartes feels, therefore I am.)

There are two kinds of "exist." Laws can exist. They do not have physical existence. Truth in general and truth of an assertion do not have physical reality, but truth can exist.

We can say the laws of physic are "real" to mean they are true. In the same sense, patterns (equals math) can be real. This kind of real, is to be true. The concept of truth is an abstraction/extraction/generalization/distillation of the observations that individual claims can be so, and Truth is noun-ified (objectified, per human nature) and bottled as a sort of thing itself. Once we declare that "truth is possible" and "truth exists," from there on we can say the more abstract of the two kinds of "exist" is about what is *true*.

So much for the existence of concepts. Now we turn to the other, the existence of things.

Whereas one cannot simply choose a reality and thereby make it so, it is, in principle, possible that consciousness- feeling- does make reality.

Because if you have a completely consistent model of a universe that could exist, how can one determine, even in principle, if it does? One could say an inanimate object could just as well interact with its objects rather than a person, but all that interaction is also within the math model, so what then makes it real?

But the consciousness, in a world where there is consciousness, is also part of that world. The world is real to everything in it, and not real to everything outside of it. This means either, that there is no absolute reality- only a set of interactions within one, which is consistent with the fall of one absolute after another in physics- or that consciousness (or other thing) exists without a need to interact in order to exist, and does not care whether you or anything else can tell that it exists.

One might say that a universe exists because it has an energy content. But the concept of energy is a codification of the rules by which everything interacts, and not something specific that can always be measured by any single kind of instrument. It is the ghost moving from manifestation to manifestation, that is, from one realness to another.

Energy applied is work, which occurs across time. Energy is about action, about change, about time.

There is no property that all particles have except energy. To have no energy is to not exist, to not be real. Energy is transactable. To exist is to be able to influence (interact). Higher energy equals more influence. Zero energy = nonexistence. Thus, the joule is a unit of existence. What do you add to mathematical laws to get the Real? Energy. Every particle that exists has energy; to have energy is to exist.

All interactions involve transfers of energy, in the space and time across which it is transferred. It could be said that one of the few things that is real is energy, but we don't know what it is. It is a mathematical construct showing conservation and force laws between a great many different happenings. Is there more to know about the physical manifestation of the energy values that we calculate?

"Energy" is a mathematical creation, which summarizes the rules of transaction during the various activities of the Universe. The concept is generated from the Laws. Yet, it is useful as the simplest single thing in reality.

Why this way- why is this or any universe the particular way it is? There is no why this exact way, it is a possibility. The Anthropic Principle also means that there are probably many other universes "somewhere."

Speculation about before a big bang has no utility as to what to do about anything or how to feel about anything, with the exception that at base everything in the universe does appear to be one in the most abstract sense. Speculation about during the Planck epoch has no utility as to what to do about anything or how to feel about anything, with the exception that at base, this *feeling* thing that we humans do and that is so fundamental to us, itself appears to be fundamental in the Universe.

Note how the nature of the Universe is not to totally constrain or not constrain, but to partially constrain- or, put another way, to constrain rather than determine or not determine. It is evolved, human, goal-driven thinking that finds this less than satisfying. The middle ground of constraint fosters emergent and high complexity, while the extremes of total or no control tend to kill or prevent it.

Let us, too, think afresh about reality by starting from ourselves. What we sense from outside without error is real. The thinking and feeling events that we sense from inside are also real. What does real mean? In our world at the least they are real. How real is our world?

The things that happened did happen; I can see this, whether anyone else can see them or not. If things exist only because they can and happen to, they still exist. Things can also possibly exist and yet not actually exist anywhere.

It could be that everything that can exist in theory, exists in reality (the Principle of Fecundity), in which case there would be no such different thing as *possible,* viewed from across all existence. But there is insufficient evidence or proof for that, no good cause to believe it.

Things may or may not be interacting somewhere. If there is a world to enter and you do (and maybe even specifically make it) then things there are real.

The fact that you can't sense it does not make it not real. And the fact that things can exist that are not sensed by some does not make everything that is possible real for someone.

Somewhere things could be interacting. There is no proof that therefore somewhere those things *are* interacting. They may not be. And because they may not be, there is a difference between the hypothetical and the real (even if all hypotheticals were to *happen* to be implemented).

Therefore, per Descartes: since some kind of interacting is happening in my head, I exist, at all, somewhere. I exist even if I am not in your universe.

Now, in my universe, things feel- at least I do. Everything that makes me feel exists (including real thoughts happening, and real imagining of unreal things).

If I can set up tight experiments resulting in a predicted feeling of mine, I can show other things exist. I then know a bunch of things that exist. I do not know if additional things exist. I also know some things that cannot exist, because they break the rules (assuming I have ascertained correct rules). A ten-pound stone that does not fall does not exist. My feelings exist. It is clear that at least some of my feelings are a consequence of other interactions in my universe. Those things also exist.

The quantum wave functions are real as rules. They are not Real. That is the simplest interpretation.

There is no proof that all the possible worlds exist, and also no need for them to. Again, it is human nature that wants a reason for everything. Nature only partially constrains, intermediate between total randomness and total specification.

Things don't all happen at once, and when they happen is also only constrained, not specified. A wave function only lasts till an event happens, and what it is is a constraining rule that makes this world possible. Various other worlds probably exist, and much less can happen in most of them. We are in one of the few that can support human life. There are probably very many. But that does not mean literally infinite. Which suggests to us that our universe itself is not infinite, just very large. In any case, theory says it has a fixed amount of matter and energy. Time started a fixed amount of time ago. Only the space dimensions are unknown as to whether now infinite, in many directions. Time could go forward forever, but not back forever.

What shall happen, has not happened yet just because it is predictable and calculable (per special relativity, in other inertial frames it may have already happened, but not in yours).

When a particle of ours is completely un-slowed it moves at c, the speed of light. It is only slowed from that if it has rest mass, or experiences other energy instead of a vacuum (and since we now know the vacuum has energy, what would c be if it did not?). (What radii do we get if we envision fermions as bosons in a tight orbit? It depends on the energy of the boson.)

Quantum collapse events could be moments of conscious feeling. Then the event and the consciousness would be the same thing. They might also be associated with some energy transfer.

Deutsch asks, if there are not many worlds, then in a quantum computer, where are all those calculations happening? Those calculations aren't "happening," we've just changed what is possible. While our human minds struggle to calculate against our human models of the actual universe, the universe does what it does, in its (semi-)orderly fashion. It neither calculates nor moves like a calculator.

Next, we deal with relativity, and the relativity of relativity.

Whether other universes exist might be a meaningless or intrinsically too-difficult question. *Inside* our own Universe (i.e., relative to other parts of it), things do exist or do not exist. Thus, the answer to whether anything is or is not truly real may be: "both."

The same is true of good and evil. Without pain and pleasure (or pain, at least) there is no such thing. Morality derives from their existence. But once they exist, good and evil exist (and our motivation to define the words is motivated by our experience of pain and pleasure in witnessing good and evil). The attempts to reduce it to purely absolute and purely relative are both in error intellectually. Assuming pain and pleasure, which we can, good and evil then absolutely exist. If nothing else, we can *define* them, once there is good and evil. Whether good and evil are "fundamental" becomes a semantic word game. There is identified a third category (which also applies to ethics), between absolute existence and relative existence. This book adopts the term "dependent existence."

Next, we check for a problem in our thinking by asking, "does a possible deep universal fabric of consciousness matter in searching for a physical theory of consciousness?" Here is what we mean by this question: Suppose that knowledge is structurally contained in this universe, or a containing multiverse of many worlds, as has been proposed by others. Imagine that our minds are simply "players" of this extant, effectively static data. Consciousness then is just an experience similar to travel along a fixed structure.

First of all, this theory is described in terms of the cognitive. It does not address feeling (it could be extended to, but that would be very much a new conjecture in such a theory, calling for its own justification and explanation.)

A reason this inquiry does not really matter, anyway, is that space and time can be presumed for the rest of physics in constructing a physical theory. Time exists, if only as a dimension, and the "fact" that we (our consciousness) just flows along it in some way does not look particularly useful or relevant to the construction of an effective physics theory for the natural phenomenon of consciousness. In this book we argue for extending existing physics and its interpretation, not unnecessarily overturning them.

Next, we move to the epistemological, and ask what it means for us to know. The ingredients are: correlation, brute fact, and experience. As we push further and further down, "the universe just works that way" is the right answer. That statement is about laws, which describe interactions. Even further down, "it just is that way," about fundamental objects appropriately describes the "bottommost turtle." In this universe some things simply are as they are, as long as they are a way that permits human life. The fundamentals should be simple, generalized and comprehensive, so that our present universe could have likely evolved from there.

We know what we feel. Given these feelings, brute facts, and laws tha we find in this anthropic universe of ours, we set about to find correlations, and show them to be consequential, second- or third-order, emergent "laws" of our universe (such as "rocks fall, in this way").

Just as with a child asking a series of Whys, asking repeatedly "and What is that?" eventually gets to "it is what it is." Feeling- fundamental consciousness- is basic, for example; it cannot be deconstructed. It is not magic, but it is elemental (and probably more so than matter; this means consciousness could even be a constituent of matter).

Other self-consistent worlds can exist in principle so they maybe they do in reality. "Why not?" we are right to ask. It is a human prejudice that *something* needs to be justified over *nothing*, given that *something* is possible.

This is particularly interesting in noting that the energy was born, and the universe continues to have a certain nonzero amount of energy. Mathematically we can justify that through quantum fluctuation. The math just serves to make the whole system consistent with what we observe today. But the whole possibility of having any energy at all turns out to be something that could be and happens to be.

There is an extremely wide set of possibilities, including the very strange to us in this world. And so, probably other universes exist "somewhere" else, most of which are likely bland and boring to our eyes. In that sense there is a multiverse (but probably not a splitting with each wave function collapse).

The wave function is a description of rules. Why does a collapse occur- what causes it? That is human thinking again, thinking there *must* be a why for everything. Our physics rules things in and things out but does not micro-dictate everything.

In places, the math speaks not of how everything is at this instant, but how fast something is moving or how fast it is accelerating, resulting in *eventuality* of position should nothing impact the locality (this builds on the concept of time or at least related difference). So, too, the universe rules what may play out and biases probabilities but does not demand a specific answer. It is this middle ground of controlled chaos that fosters the ongoing emergent evolution of things like us. The alternative to be able to do so would be a very vast deterministic universe, but both the math and simulation runs, as well as our good thinking about the problem, show that the economical solution of constraint without determination, is what is at play.

Are you determining the wave function collapses or are they determining you? At first glance it appears to be the latter. But if feeling is deeply related to wave function collapse, the matter gets more complicated and interesting. You might say feeling, too, is just another physical thing, but feeling is *you*, it is exactly the necessary and sufficient thing that separates you from a machine. This leaves room to speculate on who or what then is in charge.

If feeling occupies no equal or lower ground to wave function collapse, then *you* have no control. You do as you feel but your feelings are purely a consequence. If the reverse is true, *you* do have control. Your feelings- *you*- are part of nature, but you influence your world.

What is interesting about both wave function collapse and neurons, is that they both, from a cloudy situation of possibilities, make a distinctive *decision* to a single consistent choice. Creating hard-edged objects and numbers and words from a messy sea is a deep, fundamental aspect of consciousness.

It is also very possible that the Russian Anthropic Principle (our universe is miserable because there are more ways to be miserable than happy, so we are more likely to find ourselves in a miserable universe) is true! And one reason there are more ways to be miserable is likely that pain ("do something else, this is not working!) is more important for learning than pleasure ("this is working great!").

So, as to reality, things can be:

- Not possible (conflicts with the rules of the universe and its present state).
- Possible but not so (mathematically exists). <- not high enough reality to completely describe consciousness
- So (really exists), such as actual laws.
- Physically real (physically exists, which in our universe means it is manifested of matter and/or radiation, in interaction, revealing potential differences and forces). <- consciousness is connected to this level

(Fundamental consciousness occurs in our brains as interactions of the physically real.)

Appendix II: Brief Summary of Our Universe

All things in this universe that are real are made of particles (quanta) and changes in the relationships thereof. Because of relationship, the rules are real also (as truth). Thus, two kinds of things are real, objects and information. Information must have context to be understandable and the objects provide this. The whole universe is made of particles, interacting, in spacetime.

Consciousness, being real, is at bottom made out of particles. As with all other things that happen, feeling is probably an interaction, carried by the senton between particles with joy charge.

The existence of particles as well as their interactions are demanded but not completely described by the so-called laws (iron-clad patterns of behavior). As part of these laws, certain specific values (mostly strengths), are specified as two dozen fundamental *constants*. In universes with most other combinations of our constant values, higher consciousness would be impossible.

Our systems of physical units are based on and derived from seven classical fundamental physical quantities:

1 Time

2 Length

3 Mass

4 Temperature

5 Amount of substance

6 Electric current

7 Luminous intensity

There are also Plane angle and Solid angle, given space and time.

Time is a dimension ordering events- to us it appears to be a smooth same-rate flow but at low level there may be many spontaneous events comprising it. Anthropically, time must exist as a phenomenon in our universe, because consciousness requires time. (That fact alone does not prove time to be most fundamental, however.)

Unlike in the space dimension, there is an arrow of time: it only proceeds in the direction of higher entropy. This makes mind/consciousness a consequence, in part, of entropy.

Space is described by and is distance (length) and direction (angle).

Einstein showed with special relativity that there is relationship between space and time, put together as spacetime. With general relativity he gave us a model of gravity as distortions of spacetime.

Spacetime was created just after the theoretical Big Bang, when gravity "froze out" from the unified force.

There are additional fundamental properties of objects in the universe including spin, color charge, and weak isospin.

Quantum mechanics aka quantum physics is our description of nature, and is cognizant of:

1. Quantization: quantities restricted to discrete values

2. Wave-particle duality: objects are both like particle and waves
3. There are limits to the precision with which quantities can be known (per Heisenberg)
4. The wave function of probability amplitude

"Particles" aren't particles or waves with which we are familiar; we use particle and wave concepts to understand what "particles" are like, in terms an Earth visuomotor mind can understand.

Per Heisenberg, there are complementary variables such as position and momentum. The more precisely we measure one, the less precisely we can simultaneously measure the other.

The "ultimate (bottommost) reality" (*metareality*) is the precise probabilities of the wave function, the energy content of the universe, and two dozen constants. Probability of what? Physical values manifesting in the physical world, not just the math world. This manifesting is quantum wave function collapse.

Appendix III: Neural Module Architecture

Simulation experiments by the author have confirmed the efficacy of the below circuit strategy, dubbed "RRX." It is a form of what is now called "deep learning."

The Recognize layer is strengthened by *repetition*. This layer matures its synapse strengths earlier (they are also probably more specific on birth in their initial connections).

The Ranking layer is strengthened by *value*, which is why our brains keep them cycling given high value, so repetition will get the learning job done.

Last is the mutual-exclusion-to-one circuit (which is typically done as an extra-neural mathematical operation in artificial neural networks).

Everything is trigged by recognitions. The higher activity causes AGC (automatic gain control) and punching through to other modules that match, too.

Excitatory inputs to the Ranking layer can be used for attention control.

Appendix IV: Thalamocortical Attention Circuit

The thalamus and neocortex form a system. The thalamus is the gateway through which all external senses pass into the neocortex, except smell, which goes directly to evolutionarily older cortex. Smell has many different receptors for direct identification of all the basic and commonly-encountered smells (molecules). The other senses require analysis to understand.

The thalamus controls what input makes it to the neocortex. In this way, it controls what the neocortex is attending to. Stripped to its theoretical essentials, such a system can operate with the below architecture. This is a high-level view of circuitry that, in principle, can underlie the capability of attention.

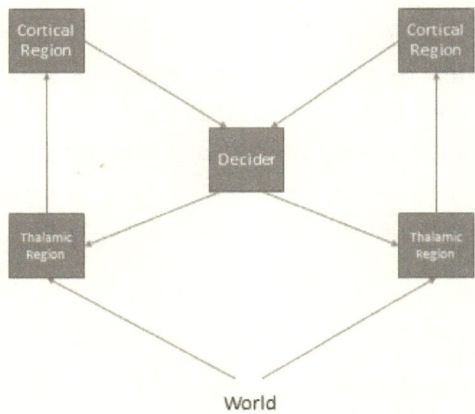

In more primitive animals, the deciding function may have rested more in the thalamus, with the more active input bundle mutually-excluding the weaker ones. As the process of attention became increasingly elaborate and volitional during evolution, the cortex took over (as with so many other functions). The thalamus now functions as a commanded gate.

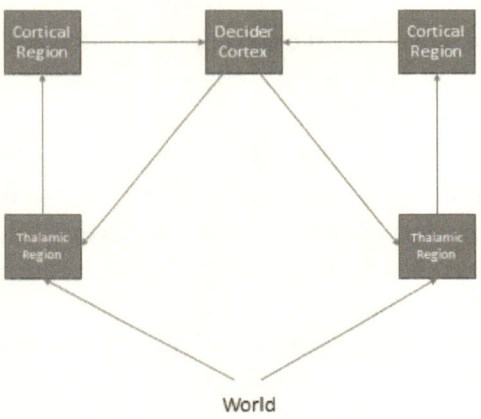

Since each chunk of thalamus is attached to its own chunk of cortex, why turn off the unimportant input at all? So that the converged upon higher-order cortex will get its input from the important source, rather than having a lot of noise "ORed" in. This is a bit like the AND-OR tree multiplexor strategy in digital electronics design.

Next, we remove the special central decider in favor of decentralized, distributed design of repetitive sameness. We realize that the brain is not arranged with a central decider (at this level) to which the thousands of cortical regions converge and from which thousands of gating signals diverge. The brain uses a decentralized system of recognition, with input from value centers, which result in some streams and networks having high incidence of recognition and value. The highest activity chunks of brain are the ones to continue processing in for the moment. Therefore, a signal back out of the chunk indicating that the input to it resulted in a very good response (or a poor one), is a sufficient indication to the thalamus of what input is worthiest of passing on.

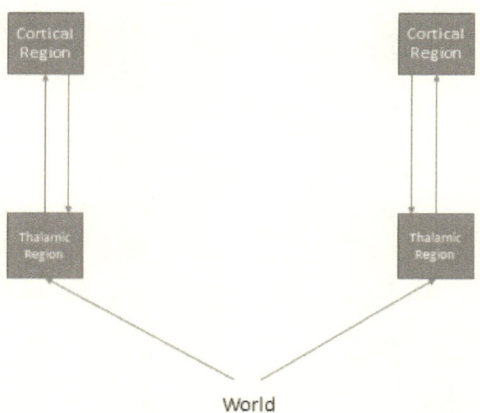

Active cortical chunks send excitatory signals to their thalamus chunks. The other thalamus chunks need to block their world input. Theoretically a lack of activity in a cortical chunk could cause feedback spikes out to the thalamus that are inhibitory. An alternative, and more likely and natural for our brain, would be that that mutual exclusion circuits exist, so that quiet signals get killed off. Most naturally these circuits would then exist in the thalamus. In principle the *cortical* circuitry that emboldened the successful chunks in the first place could discourage the others so severely that, with feedback to their thalamus chunks almost nil, no significant signal would be passed on. In either case, there is automatic gain control (mutual exclusivity).

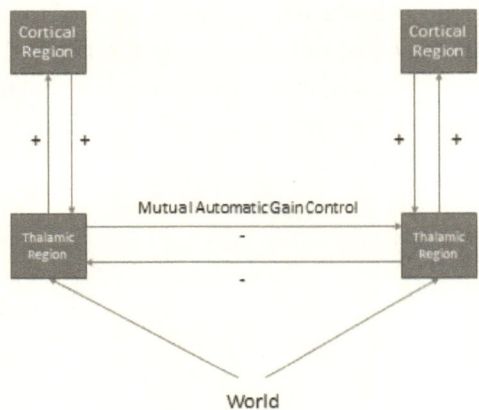

This concludes our clean-sheet-of paper theoretical design for attention control.

In the actual brain, the regenerative loops from thalamus to cortex and back are known to exist. In addition to the targeted chunk-to-chunk paths, which handle attention and involve the ventrobasal thalamus, is a diffuse set of recurrent paths, which handles wakefulness/arousal, and involves the centrolateral thalamus. Associated reticular neurons are inhibitory.

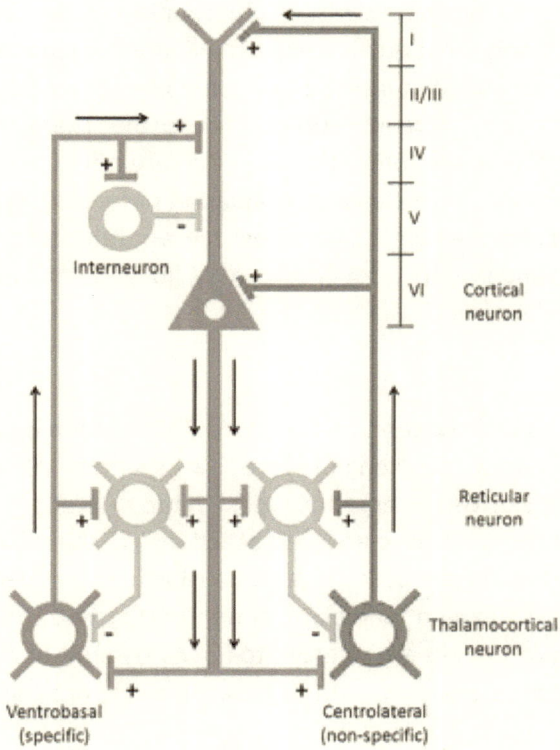

The nonspecific (non-sensory) thalamic cells' axons synapse in cortical layers I and VI; the sensory, in layer IV.

The blossoming of neural cascades that ends up capturing attention as we have described, will be caused by cognitive recognitions and emotional value, more of the former in some cases and more of the latter in others.

Quite interesting, is that this model can be extended to control patches of neocortex that are not primary sensory input areas, but those that perceive potentially important, higher level abstractions. In other words, this mechanism allows one to select not only what outside objects and phenomena to pay attention to but what inside objects and phenomena and thoughts to pay more attention to, and further examine. We even "put" matters "off to the side" that are not of prime importance. As our relevance evaluation becomes further and further elaborated and distant from raw emotions and reactive impulses and primitive training, quality "judgement" and "intuition" appear.

Two things cause attention: novelty/difference and emotion. Deliberate concentration does, too, but as with all deliberate acts, that is driven by emotion (will). In the case of concentration, we have an internal action, like the moving of a limb, that is chosen, perhaps after pausing (another deliberate act) to let judgement percolate, and then executed to affect what neural chunks are selected to flower dynamically in neocortical associations. This is a classic ability of a high-level brain.

Appendix V: Concluding Speculations

This section should be viewed separately from all the rest of the book. Here we freely explore the highly speculative. Note also that nothing in this appendix affects our thinking in the Philosophy appendix.

Humanity is at a point with consciousness similar to where it was not long ago on the topics of the electric vital force, and the chemistry of life versus "inorganic" chemistry. It turned out that electricity and chemistry are native not to life but to the universe even before life, and furthermore, made from small, uniform, and simple bits, exploited and organized by evolution. We will find the same is true for conscious feeling.

The suspicions of that time were almost right. A live animal is much different, very basically different, from a rock. While there is no vital force or different chemistry- a cell is a machine- there is a separate conscious force.

While we do not yet sufficiently understand dark matter or dark energy or a handful of other things, there is little cause for doubt that these too will fall to the same kind of mathematical thinking that has taken us so far. The Theory of *All* will include consciousness.

Repeatedly in the past, new major revelations have caused *multiple* significant mysteries to fall and to be revealed as having a common source in the newly-understood facts. Quantum Mechanics and General Relativity each solved many outstanding mysteries at once. This is not surprising, as the later pieces of a puzzle tend to bring the whole picture together. As we get down to the last pieces, predictions about them naturally present themselves.

Once it was realized that the earth is a sphere turning, orbiting, and traveling, many mysteries, from sunrise to the non-movement of the pole star, all fell into place at once, and were found to have common explanatory origin. Today, it is particularly tantalizing that the exact meaning of the collapse of the wave function and the nature of consciousness are both unanswered.

What else do we know is there but is hard to directly detect or to understand? Dark matter and dark energy. For that matter, we do not yet understand regular matter. Of the big remaining questions, it will not be surprising that several will probably fall to one stone.

There is also the matter of where the values come from for the relatively large number (over two dozen) of "fundamental" constants of the universe. That there are so many suggests that there is more to know than we do, that some of these can be derived.

These constants are

1. Fine-structure constant, characterizing the strength of the electromagnetic interaction
2. Speed of light
3. Vacuum permittivity
4. Planck constant
5. Gravitational constant
6. Electric constant
7. Elementary charge
8. 9 Yukawa couplings, giving the rest masses of the quarks and leptons
9. 2 parameters of the Higgs field potential
10. 4 parameters for the quark mixing matrix

11. 3 coupling constants for the gauge groups SU(30) x SU(2) x U(1)
12. Phase for QCD vacuum
13. 3 more Yukawa couplings and 4 lepton mixing parameters to account for neutrinos having mass
14. the Cosmological constant, for Dark Energy (0.7)

Perhaps one of these will be explained when we know more about consciousness.

Other questions: Does consciousness have any mass? It should have at least relativistic mass. What is the energy amount of a typical senton? Is there indeed consciousness traveling radiation?

What is the nature of the other dimensions that TOEs typically say exist? Are they "for" anything? Does anything happen there? If so these happenings would be invisible to our present instruments. Conscious feelings and clusters of feelings happen to be "nonextended things," having no detected size.

Probably evolution on other planets has exploited consciousness, too. And this is fundamental consciousness as we know it, because that is of the nature of the universe. This means the pain and pleasure we know are widespread, we have an understandable commonality with other beings, and good and evil may have come to exist in other places as well, and in some places, moral behavior.

Are there yet other forces that, at this late date, have still not been discovered by evolution (or by us) and exploited? For consciousness, this seems to have taken a long time, waiting for the appearance of complex animals.

Another Big Thing we don't yet know is the proper *interpretation* of quantum mechanics. This presumes humans are capable of such an understanding. It appears likely that we can, at least in large measure.

Does quantum teleportation have any connections to or similarities to consciousness, for example, by making use of the small, curled dimensions? Are non-local things local in other dimensions?

Do probability waves exist also for consciousness? They should. Is wave function collapse a local consciousness event?

Does the mystery of matter have connections to consciousness? Let us now turn to that mysterious phenomenon we call "matter." All matter particles have rest mass. Some bosons have rest mass, though (the scalar Higgs boson and the electroweak force's). What makes matter unique is that it takes up space.

Complex structure only exists (to our knowledge) where there is matter. Stable matter itself, which can form large structures, is not an early phenomenon, but only came into being a while after the Big Bang.

Matter

   uniquely occupies its space

   attaches to each other matter bits in certain ways

   stays relatively still rather than zipping past each other

Radiation does not do these things and complex organized structures are not able to be built out of it.

Matter particles effectively have size.

Matter still has its energy before its wave function collapses. A view is that matter is fully made at the collapse, making matter from probabilistic "meta-matter."

All fermions have rest mass (and therefore all fermions must travel at less than c). But unforbidden is a fermion with zero rest mass, as was once thought the case for neutrinos.

A photon can convert into a matter particle, decreasing its speed as though the primary direction of travel suddenly becomes a very tight circle (roughly similar in scale to a curled-up "extra" dimension of 11-D spacetime), that is, where its momentum splits into two vectors, one of which extends into one of these closed, compactified dimensions in which it thus orbits.

The Higgs boson, which provides rest mass, is scalar as to our familiar dimensions. How does it map to 11-D space in a correct theory?

At lowest level it may be that things blink in and out of existence, without permanence, and the permanence of objects that we observe is a macroscopic phenomenon.

Observing (interacting) locks a thing where it happens to thereby be placed then, connecting it to other things that interact with it and making its ability to be somewhere else no longer possible. Probability waves define what is possible, observation changes what is still possible.

We can see there are missing spots and missing explanations in a periodic table of particles. For example, there is no fermion with color but 0 electric charge, and no boson with color and electric charge. M-Theory predicts new particles. We know of no example of a fermion also being a force-carrying particle, but could there be one?

If we could measure the brain precisely in the dimensions with which we are familiar, would we find changes in total energy when feeling occurs?

Basic phenomena have been discovered that are not really explained, which for that reason alone might be linked to base conscious feeling events. Note how related to each other some of these are.

They include

1. Nature of time and space
2. Other dimensions and "nonextended objects"
3. Matter

4. Gravity
5. Energy
6. Wave function collapse
7. Existence and reality
8. Quantum teleportation
9. Observation
10. Feeling
11. The specific parameterizations of this universe, and reasons for the values.

The electroweak force interacts with neutrinos. Dark matter is thought to be composed of particles interacting through the weak force.

We know of two instances already of energy permeating the vacuum: dark energy and the Higgs field. Most of the energy in the universe is vacuum energy of space aka dark energy. It is conceivable that this is consciousness. The joy field might be dark energy, or something new to us.

Electromagnetic radiation (light) has two orthogonal dimensions of varying fields (electric and magnetic), both orthogonal to its direction of travel. Even in our own experienced dimensions that leaves that third dimension- the direction of travel- for another orthogonal varying field.

Much as early Man looked at the stars he was among without knowing he was made of the same stuff, Man could now be looking out into dark energy without understanding that his feeling mind is of the same stuff. The first expectation to explain a feeling of kinship when staring into the stars, is that our evolved brains and minds have created this sensation as a spurious artifact. It is possible that we don't just think we feel a connection to Nature on a large scale, but that we actually do.

Next, to be honestly thorough, we must spend a moment on human anecdotal evidence. Testimonial reporting is a notoriously weak type of evidence, but it is evidence, and is used in science and therapeutically.

Traveling much further into the speculative, then, we note that long-distance emotional connection between related individuals has been oft-reported. It is conceivable that these connections felt are real. In laboratory settings there has been no confirmed discovery of otherwise unseen connection between minds, but this could be very difficult to capture. What has often been reported are uncommon events of pain and concern between emotionally and biologically close but physically separated individuals, during occasional extreme experiences. (If the receiver is in REM sleep at the time this might occur during a dream.)

Like complaints of fibromyalgia a few years ago, these experiences are written off as autopsychogenic. In our theory, your own mind and another's are each a cloud of feeling points in a common consciousness field, and interactions may be appearing, either across our spatial dimensions or another physical dimension in which the minds are not far apart. And it would be consistent with the fact that these reported experiences are typically emotional in nature and of vague, raw emotions without cognitive descriptions.

Similar, are reports of feelings that something bad happened in a place, earlier. Fading yet persisting perturbations are conceivable. The emotion field might linger. This suggests that it somehow reinforces itself for long periods or is simply slow to decay, unless changed. (Emotions at human level do indeed have persistence- a memory of sorts- and this persistence is useful for brains in reaching conclusions and learning.)

Since the senton is a boson, it might propagate for long distances before interacting, constituting radiation- consciousness radiation. It could also exhibit quantum teleportation.

Feelings like these, and déjà vu, do repeatedly occur and do have *some* cause. There is no doubt that the dissonant *feeling* that something is wrong is itself real, and this may even have evolved as an advantage rather than be a leftover side effect. Eventually, we will be able to see them all being caused, whatever the cause is.

Appendix VI: Philosophical Note

Discoveries in physics, by definition, give us more information about how the universe works and how things happen and come to be. Some of these call for reflection as to what they mean for our plans and outlook and thinking processes. For example, quantum physics showed us it is *not* the case that everything happens just so for a reason.

Now we understand more about feeling and consciousness. What then shall we do?

Let us say that you have just awoken afresh, without any knowledge or preference of current condition. What is the top priority? To readily assess your apparent situation and select and execute whatever actions are called for your immediate survival and health and those of the people and things most important to you. Once that is done you have options, and the next question is the same question across all important time and all important things, that is, *what is the best way to live*? All other thought and action, when not operating in error, will then serve the conclusion you adopt.

Indeed, the early Greek philosophers believed the proper use of philosophy, which at that time included science, was exactly that, to understand and use the answer to that question.

Pain and pleasure are real. With the capacity for thought, humans are able to execute good and evil acts and good and evil become real. Love and hate and trust and faithlessness matter.

Matter to whom? The Universe? No. They matter to the ones asking the question, those that feel pain and pleasure: us. And those like us in the universe who we will meet, the larger us.

The rest of the universe has many trillions of glints of pain and pleasure and there is no sensible way to bring those into our equation other than to know, as first and inborn knowledge, that (net!) pain is bad, pleasure is good, and the general large principle for us is to tend heavily to the good.

There are indeed two sides to an actual force of nature. An object in motion in a bad direction will tend to stay in motion in the same direction, and so too in the positive. Aligning yourself with good will bring you to the same places as others doing the same, and the good will multiply. Pleasure is real, thought is real, ergo good is real, and good is to be desired and sought.

Selfishness is raw hedonism, and what is missing from that calculation is time. Selfishness takes, and eventually the accounts run out. Then the pendulum swings back toward equilibrium, in effect, vengefully. Unbridled selfishness is a behavior trapped at a local maximum. Without cooperation in a human network, the higher hills are never discovered.

Happily, it is natural (and literally evolved) for the human animal to be connected and empathic, and individuals who would sleaze us are too dangerous to wander our own collective.

None of this requires a particular religion or non-religion, neither does it require a coldness. Regardless of what else invisible you believe in, we all know pain and pleasure are real, and we can readily see the power of thoughtful, win-win teamwork, and the completely obvious destructiveness of some actors.

Humans have survived because of a deep drive to survive, to continue. What does last are your works, and your treatment of conscious beings, resulting in what their work is and what their treatment to others is.

While at a deep level reality may be relative to something it is not relative to you, and "my reality" is not the same as reality. Your consciousness is part of what is real, not the other way around. Wishing makes nothing so, directly. Belief and commitment over extended periods become a drive that tends to make desired things more likely.

We have learned in this book that at the first moment the torrid spot of energy was also conscious energy. Feeling exploded at the same time as what would become matter. That you are a child of the universe is even more deeply so than we previously knew. Even your feelings are a fundamental part of the universe.

What makes the body and brain is that they are *organized* bits of matter. So, too, with human-level consciousness. Individual bits of fundamental consciousness are widespread; in that sense the sun may be conscious, too, but not what we would normally call conscious. It would have a white noise background of no net effect, its "voices" individual and isolated.

What are the implications, of knowing the specific fundamental mechanisms of feeling, for how you live your life? Probably none, any more than knowing that people are animals made of atoms changes how you deal with them. But knowing that feeling is objectively and cosmically *fundamental*, regardless of the details, is meaningful for many. You are not a side effect excreted by a powerful and unfeeling universe, but a full flowering of the deepest core of feeling, energy, and laws of logic and possibility of which the universe *is*. A feeling-laden mind is not only honorable but literally essential.

Where this does cause a little more reflection, is on to what degree you wish to flow with your existing Nature, vs. more willfully wanting to change it. This decision, of course, you will also make as a result of your nature. But influenced by what you've come to know (and now you know more).

Philosophically, what is suggested here is that, where it objectively costs little, it is in concert with the rest of the universe to act consistently with your nature, rather than to imagine and idealize a nature and then completely conform to that. And change that you do wish to make, should be made in acknowledgement of the nature you are starting from today and will shape.

As to good and evil: good and evil are real because they are defined in terms of pain and pleasure, which are real to the very foundations of the Universe. Evil is creating net pain. But isn't that a consequence of following one's nature? Yes! That means some people actually are evil (not good people acting evil at the moment.) But it also means everyone whose wants you can change can be made good.

And truth is truth, whether better known or unknown.

There is a general principle that things not in concert with Nature cause stresses, wear, and risk of failure. The words in this chapter are here to encourage your impulses for harmony to swell and to subdue your impulses for conflict. "Make it" your nature to improve your nature and then follow your good nature. In actuality, what will be happening is that reading this will enhance your emotions' tendency to do as suggested here, per your emotional nature.

Good and evil both exist, but for some individuals they do not. They exist as defined in terms of one's chosen influence over the emotional fates of others. They do not exist in that truly free will does not exist; some people literally can't help themselves (equals they don't want to). But where there is ability, there is no difference between want to and shall.

People's behavior originates from their brain wiring, and that they are given. Acts of evil are beyond the control of some people. Our control and influence over bad people is for practical protection. In terms of being persuaded to believe differently, they are either able or not.

Pain and pleasure are the starting point of all respectable thinking on ethics.

At a high level, where pain in others causes pleasure in self you have evil and sadism. Where pleasure in others causes pain in self you have evil and greed and bitterness. Where pleasure in self causes pain in self you have guilt and shame. Where pain in self causes pleasure in self you have masochism.

Where a feeling witnessed causes like feeling in ourselves we have good. Consistency of feeling is positive and constructive, contradictory feeling is negative and destructive.

Embrace feeling, protect yourself, align with the good, be yourself.

www.ingramcontent.com/pod-product-compliance
Lightning Source LLC
Chambersburg PA
CBHW020653220526
45464CB00001B/418